RADON HAZARDS IN UTAH

by

Douglas A. Sprinkel and Barry J. Solomon

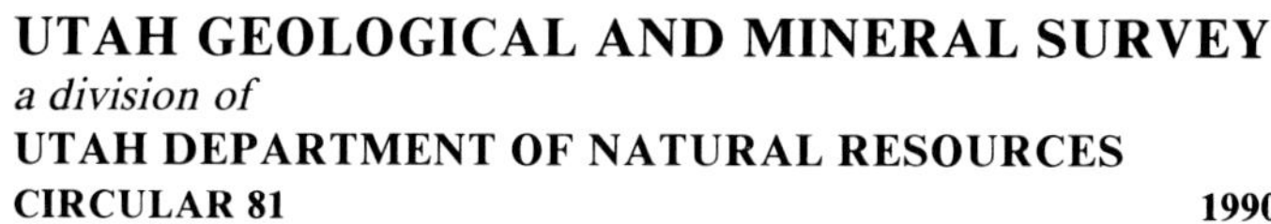

UTAH GEOLOGICAL AND MINERAL SURVEY
a division of
UTAH DEPARTMENT OF NATURAL RESOURCES
CIRCULAR 81   1990

RADON HAZARDS IN UTAH

by
Douglas A. Sprinkel and Barry J. Solomon

CONTENTS

FIGURES

TABLES

RADON HAZARDS IN UTAH

by
Douglas A. Sprinkel and Barry J. Solomon

ABSTRACT

Radon is a naturally occurring gas derived from geologic materials. When inhaled, radon decay products are a significant cause of lung cancer. High levels of radon gas in uranium mines have long been recognized as a health hazard to miners, but the hazard from indoor accumulation of radon gas at lower levels has only recently been recognized.

Geologic factors were used to identify potential radon-hazard areas in Utah by mapping the distribution of: (1) possible point sources for radon, including known uranium occurrences, and (2) generalized sources including uranium-enriched rocks (granite, metamorphic rocks, black shales, and some volcanic rocks) found at the surface or beneath well-drained, porous, and permeable soils, and soils derived from uranium-enriched rocks. The Utah Bureau of Radiation Control conducted a survey to assess indoor radon levels statewide. These levels were then compared with potential radon-hazard areas to test the utility of regional geologic evaluations as a tool for predicting where elevated indoor radon levels may occur.

Results of the study show a geometric mean (GM) of 1.8 pCi/l, an average (AM) of 2.7 pCi/l, and a maximum of 68.2 pCi/l. This compares with an estimate of indoor radon concentrations in the United States of 0.9 pCi/l (GM) and 1.6 pCi/l (AM). At least two radon levels >10 pCi/l were recorded in and near each of four areas tested: Monroe, Sevier County; Provo, Utah County; Sandy, Salt Lake County; and Ogden, Weber County. The Ogden, Sandy, and Provo areas are close to a mountain front and are underlain by Quaternary lakebeds and alluvial fans derived from metamorphic rocks, granitic rocks, and black shales, respectively. Each area has deeper ground water, and more permeable soils, than adjacent valley locations. The Monroe area is underlain by Tertiary volcanic rocks and well-drained, permeable alluvium. Thus, all four areas have uranium-enriched source rocks, permeable soils to allow migration of radon gas, and a lack of shallow ground water which might inhibit radon migration. Each area had been identified from geologic studies as having a potential radon hazard, indicating that regional geologic evaluations are useful tools to identify areas of elevated indoor radon levels in Utah.

INTRODUCTION

Most geologic hazards are natural, dynamic, earth processes that tend to alter the landscape and adversely affect the works of society. During the past decade, Utah has been subjected to such geologic hazards as debris flows, debris floods, landslides, and the rapid rise of Great Salt Lake, which together cost the citizens of Utah hundreds of millions of dollars (Austin, 1988). These hazards are governed by regional and local geologic setting. The occurrence of high radon concentrations in buildings, although not a process of landscape alteration, is now recognized as another hazard controlled by geologic factors. Radon is a radioactive gas of geologic origin, once thought of as an occupational health hazard among underground uranium miners. Radon has now been found in many buildings throughout the United States in sufficient concentrations to represent a health hazard to building occupants. The U.S. Environmental Protection Agency (EPA) estimates that from 8,000 to 40,000 Americans will die each year from lung cancer caused by long-term radon inhalation (Schmidt and others, 1990). Concern for the health consequences associated with long-term exposure to elevated indoor radon levels has prompted scientists and health officials at both the national and state levels to assess the radon hazard and to determine with more precision the extent of the problem.

Everyone receives some low-level radiation from naturally occurring radioactive isotopes present in nearly all rocks, soils, and water. We are also subjected to a certain amount of cosmic radiation that penetrates the earth's protective atmosphere. The amount and distribution of terrestrial and cosmic radiation varies with altitude and location, but daily doses of natural radiation pose a low health threat to the general population. However, terrestrial concentrations of radioactive isotopes are not uniformly distributed in rocks and soils. Some areas have elevated levels of radioactivity due to the geologic concentration of radioactive isotopes. Scientists have discovered elevated natural radiation levels in many parts of the world from measurements taken to monitor background radiation levels near nuclear power plants (Nero, 1986). Concern of the scientific community grew over the potential consequences of exposures to elevated levels of naturally occurring radioactive isotopes.

Discussions of the health effects of natural radiation began in the 1960s and have continued into the 1990s (Adams and Lowder, 1964; Adams and others, 1972; Gesell and Lowder, 1980; Vohra and others, 1982; Schoenberg and others, 1990). Increased awareness of a potential health risk from exposure to elevated indoor radon levels began in the mid-1970s as a result of research conducted in Sweden (Swedjemark, 1980). Potential health risks were associated with building sites on uranium or uraniferous phosphate mill tailings, and with the use of uranium tailings as fill material (National Council on Radiation Protection and Measurements; NCRP, 1984a). Still, most health concerns for the general population were focused on the potential exposure to radiation generated from nuclear power plants.

Scientists recently discovered that certain rock types significantly contribute to elevated indoor radon levels. In 1984, a worker at the Limerick Nuclear Power Plant in Pennsylvania repeatedly set off radiation alarms in the plant (Nero, 1986). The radiation source was found to be his radon-contaminated home in Boyertown, Pennsylvania; the home has one of the highest indoor radon levels recorded in the United States. This area of Pennsylvania is within the Reading Prong geologic province, which consists of metamorphic rocks with above-average uranium concentrations. These rocks were the source of the radon found in the worker's home (Smith and others, 1987). This revelation established the relationship between geology and indoor radon levels, and it prompted scientists to re-examine similar geologic areas.

Investigators have long known that certain rock types typically contain above-average uranium concentrations (Phair and Gottfried, 1964; Richardson, 1964; Rogers, 1964; Heier and Carter, 1964; Otton, 1988). These rocks are a primary source of naturally occurring radon gas. Based on preliminary work conducted in some states, the EPA has suggested (press releases, August 1986 and August 1987) that areas of the United States underlain by certain rock types (metamorphic rocks, granites, and black shales) have a greater likelihood of elevated indoor radon levels than areas underlain by other rock types (figure 1). However, rock type alone isn't always an indicator of elevated indoor radon levels. Other geologic considerations such as soil permeability and porosity, the degree of water saturation, and ground-water flow direction play important roles in determining probable hazard areas. Non-geologic considerations such as weather conditions, building construction techniques, construction materials, and life styles also directly influence indoor radon levels. Developing an understanding of the geologic and non-geologic components that affect the production and concentration of radon gas will significantly contribute to an increased ability to identify those areas of Utah most likely to have elevated indoor radon levels.

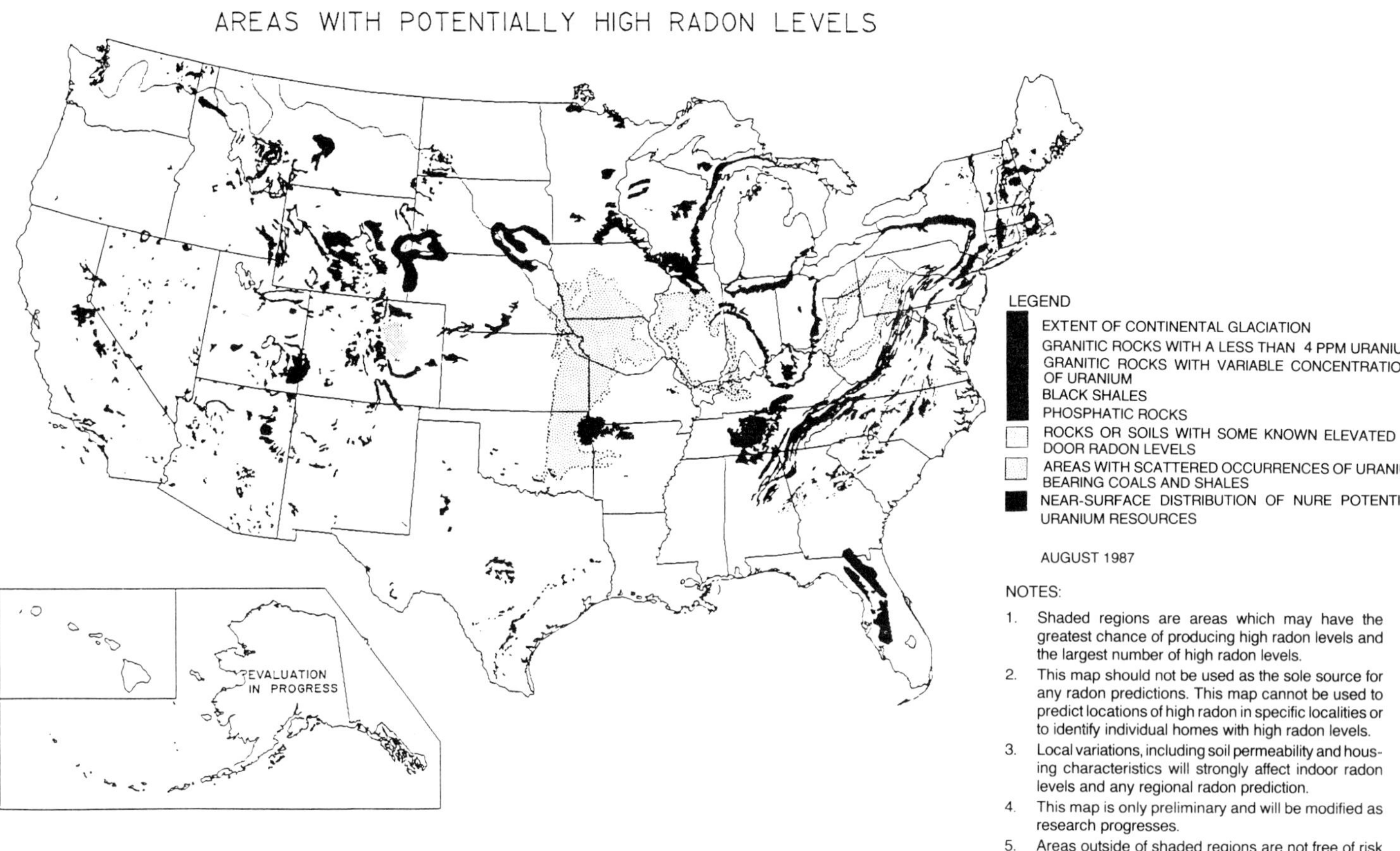

Figure 1. *Areas in the United States the U.S. Environmental Protection Agency identifies with potential high radon levels. These areas delineate certain rock types found throughout the U.S. that have the capability of producing greater than average amounts of radon (EPA, press release August 1986 and August 1987).*

Two separate strategies guide investigators in their attempt to determine the magnitude of the potential radon hazard in Utah. One is to determine the distribution and magnitude of elevated indoor radon levels through testing in existing buildings. The other is to make geologic observations and develop methods to assess the likelihood of radon hazards at sites prior to construction. Data from the first technique is needed to develop and verify the second. Information gained from both approaches will supplement one another and provide a clearer picture of the radon hazard in Utah.

Until recently, little was known about indoor radon in Utah. Indoor radon measurements made over the past few years in limited areas of the state suggested that certain locations in Utah may be susceptible to elevated radon levels (Woolf, 1987; Lafavore, 1987). Other studies (Rogers, 1956, 1958; Tanner, 1964; Horton, 1985) have addressed Utah's outdoor radon occurrences in soil and water. A coordinated statewide effort was initiated by the Utah Geological and Mineral Survey (UGMS) to identify and map rock types that are believed to produce radon in elevated quantities (Sprinkel, 1987, 1988). The results of this work guided a year-long indoor radon study conducted by the Utah Bureau of Radiation Control (UBRC) in 1988. The results of that study were recently summarized (Sprinkel and others, 1989), and are expanded upon here.

RADON AS A HAZARD

Radon is an odorless, tasteless, and colorless radioactive gas which forms as a product in three radioactive decay series. The most common of these is the uranium decay series where uranium (^{238}U) decays to form stable lead (^{206}Pb) (figure 2). New isotopes form through spontaneous disintegration and emit alpha, beta, and gamma radiation. Radon (^{222}Rn), one such isotope, forms directly from the disintegration of radium (^{226}Ra). As the radioactive decay process continues, a sequence of short-lived radon progeny form that emit mostly alpha and beta radiation (figure 2). Two other isotopes of radon (^{219}Rn and ^{220}Rn) also occur in nature and may contribute to the indoor radon problem, but ^{222}Rn is the most abundant of the radioactive radon isotopes, has the longest half-life (3.825 days), and is considered the most significant contributor to the indoor radon hazard. Subsequent references to radon imply ^{222}Rn derived from the ^{238}U decay chain.

In nature, radon is found in nearly all rocks and soils in small concentrations. Most sources of radiation are solids. However, radon is an inert gas that is very mobile. Therefore, radon can move with the air or, if dissolved in water, migrate through cracks and other open spaces in rocks and soils. Radon normally escapes into the atmosphere in small concentrations. However, large concentrations of radon may exist when favorable geologic conditions are present.

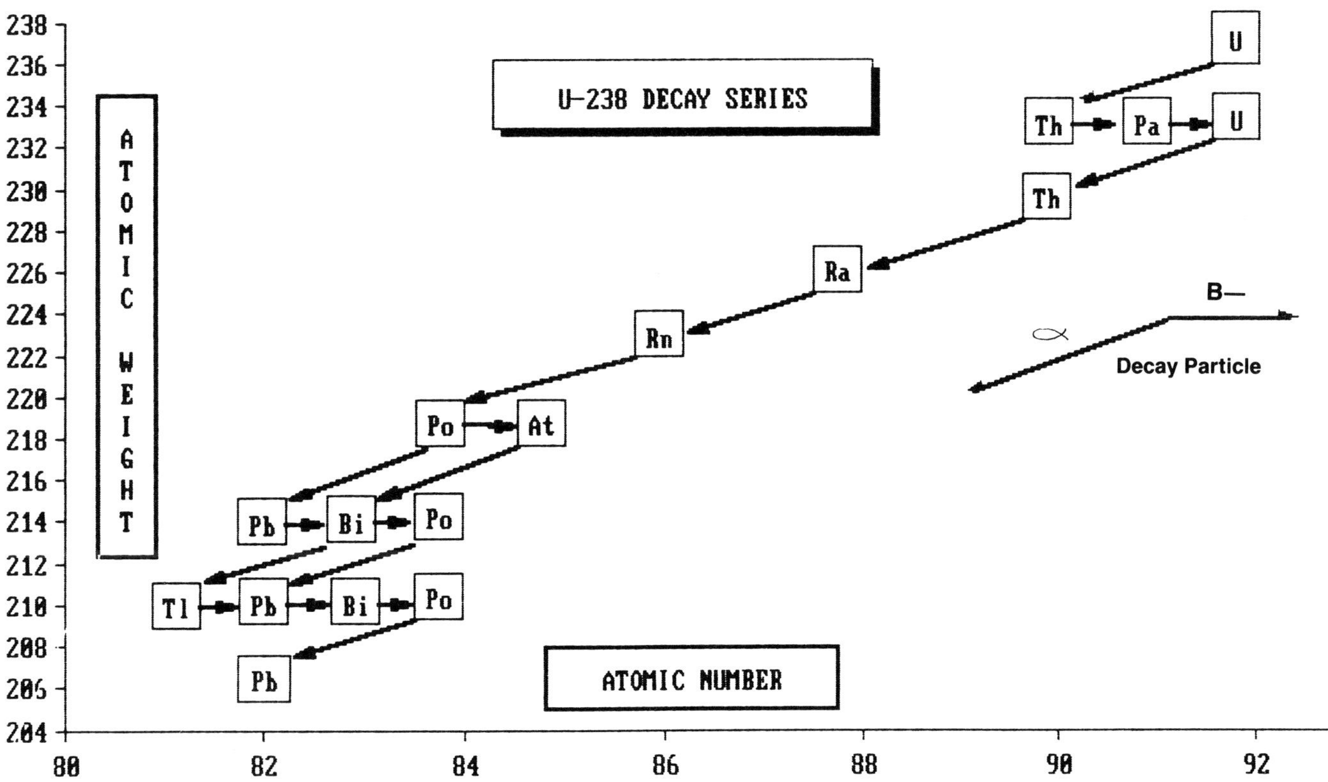

Figure 2. *Uranium (^{238}U) decay series. Radon (^{222}Rn) is derived from radium (^{226}Ra) and is the only isotope in the series that is a gas. Because it is also inert, radon has the ability to move with air or water without participating in chemical reactions (modified from Durrance, 1986).*

Because radon is derived from geologic materials, geology influences the local concentration, release, and migration of radon. Radon and other sources of natural radiation are ubiquitous in small concentrations, but most natural background radiation is of a low-level dosage not considered to be a general health threat. However, health officials believe that breathing elevated levels of radon over time increases a person's risk of lung cancer because of internal radiation damage to the lungs from decaying radon and radon progeny (Jacobi and Eisfeld, 1982; NCRP, 1984a, 1984b; Samet, 1989).

Radon concentrations in the atmosphere never reach dangerous levels because air movement dissipates the radon. People are subjected to a radon hazard in buildings or in natural enclosures with poor air circulation. The exposure to the hazard, in most cases, depends on non-geologic factors such as foundation condition, building ventilation, construction material, and life styles. Radon can find its way into buildings through small basement cracks or other foundation penetrations such as utility pipes (figure 3). Maximum radon concentrations are often found in basements or low crawl spaces (Fleischer and others, 1982) because these parts of a house are in contact with the ground, which is the primary source of radon. Radon concentration is measured in picocuries per liter of air (pCi/l); a picocurie is the decay of about 2 radon atoms per minute. Most buildings throughout the United States contain some radon, but concentrations are usually less than 3 pCi/l (Nero and others, 1986). The average indoor-radon concentration (figure 4) is about 1 pCi/l (Sextro, 1988). Long-term exposure to these levels is generally considered a small health risk to the general population; larger concentrations pose greater risk (figure 5).

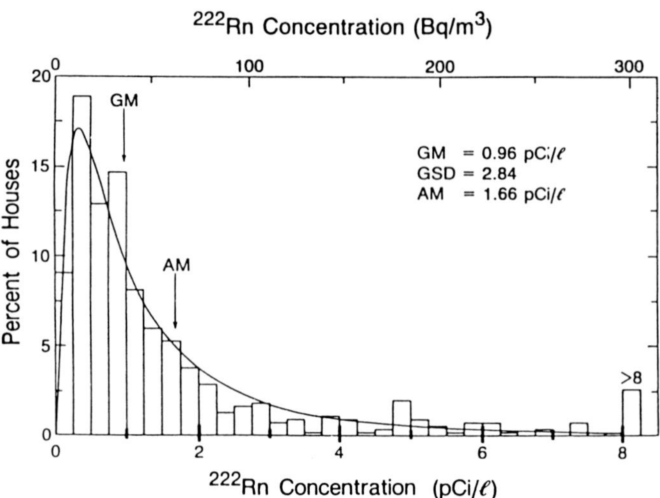

Figure 4. *The actual distribution of radon concentrations in the U.S is unknown, but this frequency distribution estimates the probable distribution of* ^{222}*Rn concentrations based on 552 U.S. homes surveyed. The smooth curve is a lognormal function with the parameters shown. The geometric mean (GM) is about 0.9 pCi/1, the geometric standard deviation (GSD) is 2.8, and the average (AM) is 1.6 pCi/1 (from Sextro, 1988).*

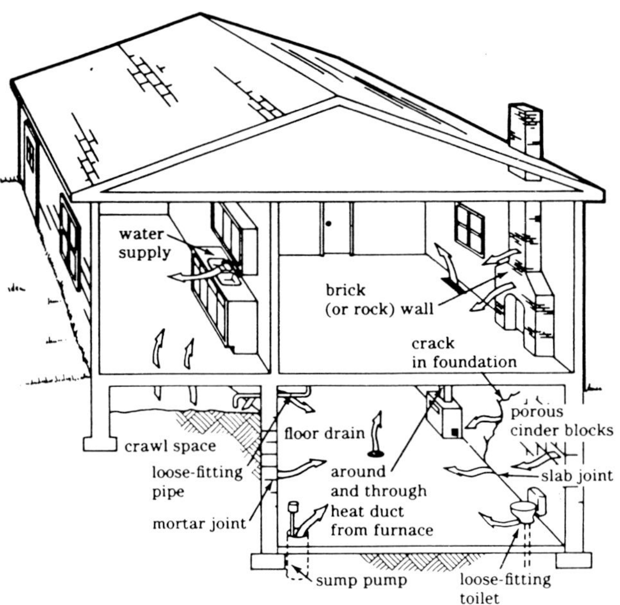

Figure 3. *Various pathways for radon to enter a home. Most of the entry routes are in the basement, because that is the part of the house with the greatest surface area exposed to the surrounding soil (reprinted from* Radon: The Invisible Threat *by Michael Lafavore. Permission granted by Rodale Press, Inc., Emmaus, PA 18049).*

Radon Risk Evaluation Chart

pCi/l	WL	Estimated number of lung cancer deaths due to radon exposure (out of 1000)	Comparable exposure levels		Comparable risk
200	1	440—770	1000 times average outdoor level		More than 60 times non-smoker risk
					4 pack-a-day smoker
100	0.5	270—630	100 times average indoor level		
					20,000 chest x-rays per year
40	0.2	120—380			
20	0.1	60—210	100 times average outdoor level		2 pack-a-day smoker
					1 pack-a-day smoker
10	0.05	30—120	10 times average indoor level		
					5 times non-smoker risk
4	0.02	13—50			
			10 times average outdoor level		200 chest x-rays per year
2	0.01	7—30			
					Non-smoker risk of dying from lung cancer
1	0.005	3—13	Average indoor level		
0.2	0.001	1—3	Average outdoor level		20 chest x-rays per year

EPA, 1986 OPA 86-004 Fig. 6

Figure 5. *Radon risk evaluation chart. The EPA (1986a) has developed this chart to provide comparable risks for people to evaluate their personal risk to the radon hazard. Units of measurement often used to report radon decay product concentrations are working levels (WL), noted in the second column. One working level (WL) is defined as the quantity of short-lived radon decay products that will result in 1.3 x 10⁻⁵ Mev (million electron volts) of potential alpha energy per liter of air (EPA, 1987).*

Inhalation of radon is not thought to be the primary source of internal radiation because radon does not attach itself to the lining of the lungs. In addition, most radon atoms are exhaled before they decay and emit dangerous alpha particles to lung tissue. The radioactive isotopes formed from radon decay are of more concern because they are not inert and most readily attach themselves to the first charged surface they come in contact with, typically, dust or smoke in the air. People who smoke place the occupants of the building at greater risk because the smoke places a greater percentage of particles in the air, to which radon progeny become attached and are then inhaled into the lungs (National Research Council, 1988).

The dust or smoke particles with radon progeny attached become lodged in the lining of the lungs. Once lodged, the resident time in the lungs for these particles is greater than the cumulative half-life of the radon progeny. This allows tissue to be directly bombarded by a series of energetic alpha particles as the radon progeny decay (table 1).

Isotope	Symbol	Half-Life	Decay Particle	Energy (MeV)
Uranium	U-238	4.468 billion years	a	4.195 4.14
Thorium	Th-234	24.1 days	b	0.192 0.10
Protactinium	Pa-234m	1.18 minutes	b	2.31
	Pa-234	6.7 hours	b	2.3
Uranium	U-234	248,000 years	a	4.768 4.717
Thorium	Th-230	80,000 years	a	4.682 4.615
Radium	Ra-226	1602 years	a	4.78 4.59
Radon	Rn-222	3.825 days	a	4.586
Polonium	Po-218	3.05 seconds	a, b	6.0
Astatine	At-218	2 seconds	a	6.7 6.65
Lead	Pb-214	26.8 minutes	b	0.7 1.03
Bismuth	Bi-214	19.7 minutes	a, b	a=5.5 b=3.2
Polonium	Po-214	0.000164 seconds	a	7.68
Thallium	Tl-210	1.32 minutes	b	5.43
Lead	Pb-210	22.3 years	b	0.015 0.061
Bismuth	Bi-210	5.02 days	a, b	a=4.7 b=1.16
Polonium	Po-210	138.3 days	a	5.3
Lead	Pb-206			

Table 1. Uranium decay series showing the half-lives of isotopes. Radon's half-life is less than four days and the radon progeny combined half-life is about 90 minutes. a=alpha; b=beta

Inhalation of radon and radon decay progeny was suspected as a health problem in the late 1950s and early 1960s when investigations were conducted on miners who worked in underground uranium mines. The studies concluded that high concentrations of radon found in underground uranium mines contributed to an increased incidence of lung cancer among miners (NCRP, 1984b). Indoor radon problems were also believed to have been associated with homes built on uranium mill tailings (NCRP, 1984a) or uraniferous phosphate processing waste. The lower concentrations of uranium found in most rocks were assumed not to contribute to significant levels of radon indoors. The demonstration, in 1984, of an association between elevated indoor radon levels and lower concentrations of uranium found in various rocks near Boyertown, Pennsylvania, was therefore surprising. The potential for elevated levels of indoor radon is now associated with rock types having average uranium concentrations less than 15 ppm (parts per million) (Durrance, 1986). Many areas of the country, including much of Utah, are underlain by rock which could produce elevated indoor radon levels.

Changes in building practices over the past 15 years have also contributed to the radon problem. Since the 1973 oil embargo, conservation of our non-renewable energy resources has been a national goal through energy-efficient practices. The building industry has made structures more energy efficient, but they have not improved ventilation systems to accommodate restricted natural air flow. Buildings, including single-family homes, constructed before 1973 often did not use energy-efficient measures, allowing indoor air to escape through above-grade joints and uninsulated walls and attics. Today, more energy-efficient homes and other buildings prevent the loss of indoor air to the outside. Studies (Fleischer and others, 1982; Nero and others, 1982) have shown that newer, energy-efficient buildings with under-designed ventilation systems generally have higher indoor radon levels compared with older, conventional buildings.

MEASUREMENT OF INDOOR RADON LEVELS

Because non-geologic factors influence indoor radon concentrations, radon levels in buildings must be measured to determine if problems exist. Radon can be measured with both short-term and long-term passive detectors and electronic instruments. Some may be placed by the homeowner, while others require professional installation. Most people want information quickly, so they often select short-term monitoring methods which give quick, accurate results. A short-term measurement is one conducted for a period of less than three months (Ronca-Battista, 1988). However, long-term monitoring, typically for a twelve-month period, provides more realistic information.

Measurements taken over a few days or on a single day will provide only a snapshot of indoor radon levels for that particular time. Radon emissions from the ground, and resultant indoor radon levels, fluctuate daily, weekly, and monthly because of atmospheric changes (Kramer and others, 1964; Schery and Gaeddert, 1982). In addition, concentrations fluctuate seasonally because building ventilation is less in winter than in summer, and indoor heating and air conditioning affect concentrations. A longer period of monitoring is recommended to smooth out short-term fluctuations. This will provide a more realistic picture of the yearly average indoor radon concentration. The UBRC in Salt Lake City provides information on types of radon detectors available, their advantages and disadvantages, and comparative cost.

Radon measurement protocols suggested by the EPA attempt to assure accuracy and consistency of data (Ronca-Battista, 1988).

The protocols were developed to balance the need to obtain results quickly with the need to acquire measurements which best reflect long-term indoor radon levels. To accurately determine the indoor radon levels throughout the home, long-term monitoring is needed on each floor. EPA (1986b) and Ronca-Battista (1988) suggest, however, that a short-term screening measurement which follows EPA protocol (closed-house conditions) may be conducted in the lowest livable area of the house to determine if additional testing is necessary. According to EPA (1986b), additional testing is not needed if the short-term screening measurement is less than 4 pCi/l and, although a small health risk is present, remediation is unnecessary. If a result is greater than 4 pCi/l and less than 20 pCi/l, a follow-up test of a 12-month measurement in two living areas of the house is recommended by EPA (1986b). If retesting confirms screening measurements, mitigation may be warranted in a few years. If a screening measurement is greater than 20 pCi/l and less than 200 pCi/l, retesting is recommended in two living areas of the house for no more than three months (EPA, 1986b). If a screening measurement is confirmed, remediation should take place within the next several months. If a screening measurement is over 200 pCi/l, retest immediately in at least two living areas of the house (EPA, 1986b). If confirmed, remedial action should commence within several weeks. Thus, current EPA measurement protocols emphasize immediate short-term, follow-up testing in two living areas of homes with screening measurements greater than 20 pCi/l (Ronca-Battista, 1988). The UBRC follows these guidelines but emphasizes the value in long-term monitoring (D. Finerfrock, oral communication, 1987).

GEOLOGIC CONSIDERATIONS

Tanner (1986) suggests four prerequisites to elevated indoor radon concentrations. The home must (1) be built on ground that contains radium, (2) have underlying soils that promote easy movement of radon, (3) have porous building materials or openings below grade, and (4) have a lower atmospheric pressure inside than outside. The ground must contain a certain amount of uranium from which radon emanates and the radon must travel easily through the soil to the structure before it decays. The structure must have foundation cracks or spaces in contact with the ground and have an atmospheric pressure lower inside than outside to allow radon to enter. Domestic water and home construction materials also contribute to indoor radon levels, but the major contributor in most cases is the geologic material immediately underlying the home.

The first geologic consideration in evaluating a radon hazard is the distribution of rocks that may contain uranium in unusually high concentrations. Areas underlain by rock such as granite, metamorphic rocks, some volcanic rocks, and black, organic-rich shales (plus other sedimentary units derived from uranium-enriched source rocks) are generally associated with an indoor radon hazard. If the radioactive source rock is present in the ground, there are several geologic considerations that enhance or impede radon emanation and movement. Most of these factors are observable and measurable in the field. Many of the principles and techniques used to detect radon emanation and migration were first developed for uranium exploration during the uranium boom three decades ago (International Atomic Energy Agency, 1976). Radon hazard assessment uses the same principles and techniques, but different levels of sensitivity.

Once uranium is present in the mineral matter of the rock or soil, the radon formed must escape the crystal structure or surface film of the mineral grain. It does so during the spontaneous decay of radium, which emits alpha particles and radon atoms. The radon atoms recoil in the opposite direction of the alpha particles. Radon atoms near the grain's surface may move into the pore space or burrow into an adjacent mineral grain (figure 6). Because the newly produced radon atom has a small recoil distance, grain size, pore size, porosity, and moisture content are important components in radon emanating power (Tanner, 1964, 1980; Barretto, 1975). The sorption or precipitation of uranium in association with metal oxides also reportedly enhances radon emanation in rocks and soils (Gunderson, 1990). Emanating power is defined as the fraction of radon atoms that escape from the solid where they were formed (Tanner, 1980).

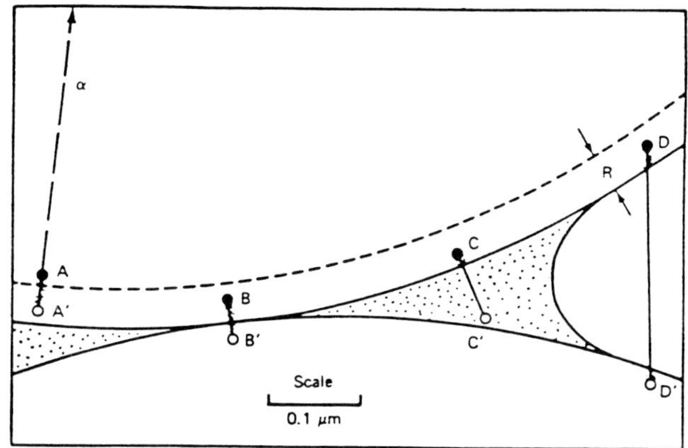

Figure 6. *Idealized cross section of two mineral grains showing how radon can escape (the emanation process). The two grains are in contact near B. The stippled pattern represents a meniscus film of water between grains. The white area to the right of the water is air. ^{226}Ra atoms are represented by the solid dots and ^{222}Rn atoms are the open circles. R is the recoil distance of the newly formed radon atom. Because of the small recoil distance of radon within the grain, only radium atoms found near the grain's surface would contribute to radon emanation. Recoiling radon atoms passing through a film of water are more likely to remain in the pore space, while radon atoms that pass only through air may become embedded in the adjoining grain and rendered harmless (from Tanner, 1980).*

Grain size and emanating power are inversely related (Tanner, 1964, 1980; Barretto, 1975). Grains larger than 1 micron can retard radon recoil because the recoil distance is less than the grain size and radon atoms produced deep in the grain's interior are unlikely to escape. Only radon atoms near the grain's surface have the opportunity to escape, thus reducing the amount of available radon atoms. Smaller grains also have a larger ratio of surface area to volume, which increases the relative amount of surface area available for the escape of radon atoms. Small pore size, though, can reduce emanating power because the recoiling radon can pass through the pore space and become embedded in the adjacent grain (Tanner, 1980).

Another factor that influences radon production is the water that occupies the space between the grains. A water coating on the grains can increase radon emanation (Tanner, 1980). When radon recoils from a grain in a dry environment it can pass through the dry pore space and become embedded in the adjoining grain. However, if the grain has a thin coating of water, the water absorbs the recoil energy of the radon atom and the radon will more likely be retained in the pore space. Water doesn't increase the rate of radon production but does allow a higher percentage of recoiling radon atoms to remain in the pore space.

Once free radon is present in the pore space of rock or soil, it can begin to move. Radon migration results from two mechanisms, diffusion and mass transport. Diffusion is the process of random movement of radon atoms by natural vibration. Mass transport is the process of convective flow of soil gas caused by air pressure differences within the soil, or between the soil and atmosphere, or between the soil and the foundation of a structure. Air pressure differences can be caused by barometric pressure changes in the atmosphere, wind blowing across a surface, or thermal convection generated by heating or cooling. These processes affect the release of radon from the soil, as well as the radon level within a structure. Home heating and wind conditions can create low atmospheric pressure inside a home, allowing it to act as a pump which draws in underlying radon-laden soil gas.

Radon was once thought to move through the rock or soil column by the process of diffusion. However, Baretto (1975) suggests that the distance radon can travel by diffusion in about four days, the effective radon half-life, is negligible. Recent investigations (Clements and Wilkening, 1974; Tanner, 1980) suggest that both diffusion and convective flow are active in radon migration. Because high radon concentrations in some areas cannot be explained by diffusion alone, mass transport of radon by the convective flow of soil gas is thought to be the primary mechanism that moves large quantities of radon through the ground (Tanner, 1964). Diffusion, however, may be the dominant mechanism of radon movement in soils with low average permeability (Tanner, 1990). Once soil gas reaches the backfill-and-subslab zone just outside the foundation, pressure-driven convective flow of radon-bearing soil gas is commonly accepted as the dominant mechanism to move radon from outside house foundations to inside the structure.

Water saturation of soil or rock columns can effectively inhibit radon migration. A small quantity of water increases radon emanation, but too much water restricts radon migration by reducing diffusion and blocking the flow of soil gas (Tanner, 1980). Radon may move with the water, but the flow of water through soil and rocks is usually much slower. Water does, though, provide an effective means to carry radon from its rock source (Tanner, 1980). Where domestic water sources contain high levels of radon, they may contribute to indoor radon levels (Vitz, 1989). Estimates of the contribution of radon in water to airborne radon range from 1 to 2.5 pCi/l in air for every 10,000 pCi/l in water (Cross and others, 1985; Pritchard, 1987). Thermal waters and their deposits (tufa) are also likely sources of radon.

The permeability and porosity of the rock or soil column also influence radon's ability to migrate to the surface. There is a correlation between areas that have permeable soils which contain open pathways enabling the migration of soil gas, and elevated indoor radon concentrations (Tanner, 1980; Schery and Siegel, 1986; Otton and Duval, 1990). Indices have been devised, such as the radon source potential (Sextro and others, 1989) and the Radon Index Number (Kunz and others, 1989), that attempt to predict indoor radon levels from soil permeability and soil gas radon concentrations. While such indices may work in relatively homogenous soils, spatial variations in most soils are large, as are temporal variations of soil gas radon concentration, making site characterization measurements difficult without an extensive sampling network (Sextro and others, 1989).

Faults and fractures are zones of rock breakage which contain openings where air and water can move. Uranium in ground water is often deposited and concentrated in such zones. However, even if uranium mineralization does not significantly occur, fracture zones may enhance radon concentrations in soil gas adjacent to the fractures by providing permeable and porous pathways for radon-bearing gas to migrate towards the surface. Measuring radon concentrations over large areas can identify these zones. Monitoring changes in radon concentrations on active fault zones, such as the San Andreas fault zone in California, or in volcanically active areas may serve as a possible indicator of future geologic activity such as earthquakes or volcanic eruptions (Tanner, 1980; King, 1986; Teng and Lang, 1986; Thomas and Cuff, 1986).

POTENTIAL RADON-HAZARD AREAS IN UTAH

There are several areas in Utah that have the geologic conditions required to produce a radon hazard. Sprinkel (1987), using regional geologic data, mapped potential radon-hazard areas in Utah. These areas were identified by known uranium occurrences (possible point sources for radon); uranium-enriched rocks (generalized sources) at the surface or beneath well-drained, porous and permeable soils; anomalous surficial uranium concentrations; and the surface trace of the Wasatch fault zone. Uranium occurrences have been previously described by Hintze (1967), Doelling (1969), Chenoweth (1975), Silver and others (1980), Gurgel (1983), and Steven and Morris (1984). Included are uranium mines, uranium mill sites, and geothermal areas. Uranium-enriched rocks have been described by Durrance (1986), and their distribution in Utah (as well as the distribution of other rock types) were mapped by Hintze (1980). A map of apparent surface concentration of uranium determined by airborne surveys (Duval and others, 1989) outlines the distribution of uraniferous rocks not otherwise shown by geologic mapping. The Wasatch fault zone (Davis, 1983a, 1983b, 1985; Scott and Shroba, 1985; Personius, 1988; Machette, 1989; Personius and Scott, 1990; Nelson and Personius, in press) is another area of Utah which is a likely candidate for producing a radon hazard. Sprinkel (1987) did not include Quaternary units in the compilation unless documented in publications to be a radon source (Steven and Morris, 1984).

Areas in Utah with a greater potential for elevated indoor radon levels, based on geologic data, are shown on figure 7. The map is only a guide to help State health officials, interested decision-makers, developers, and the public determine areas for indoor

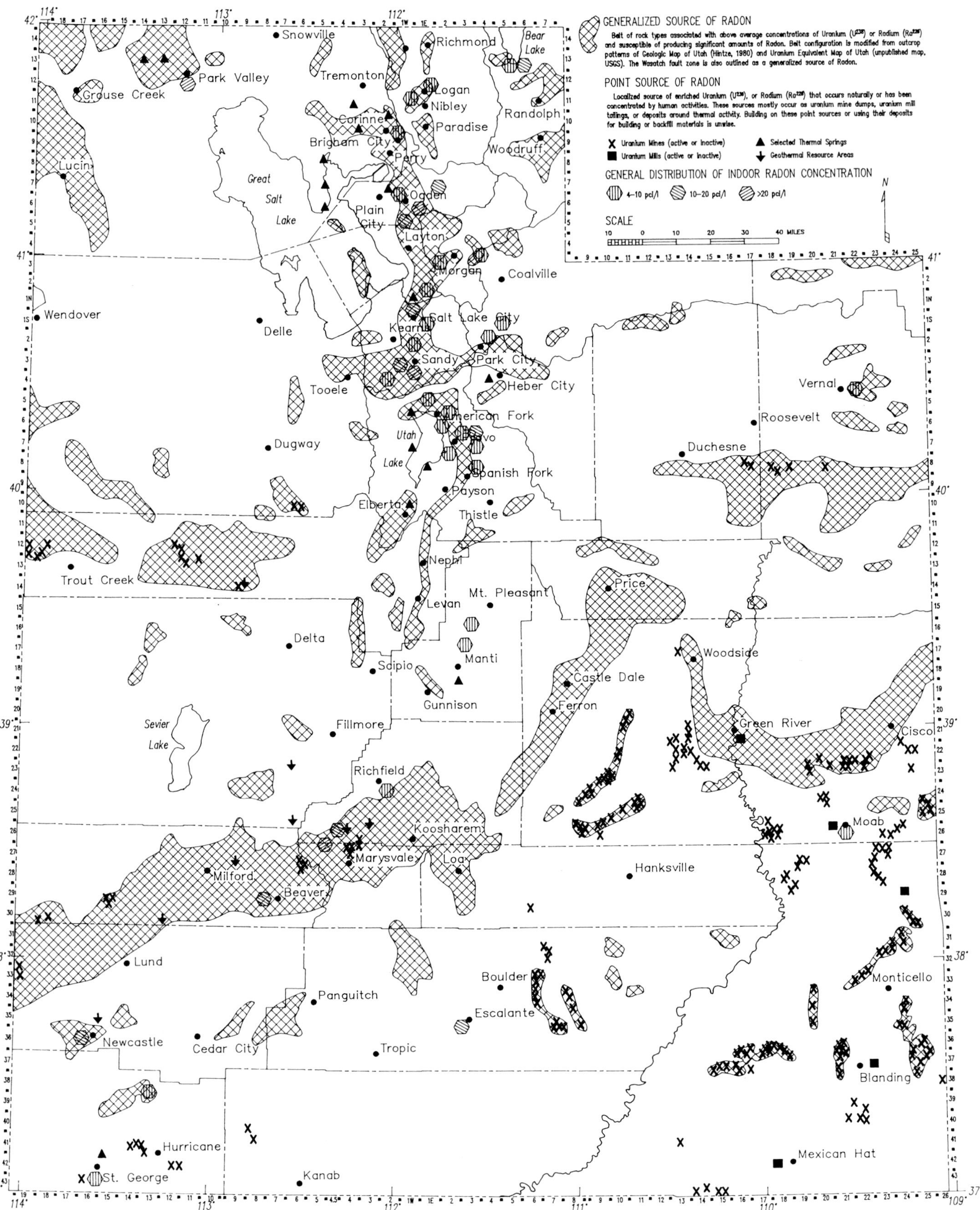

Figure 7. *Generalized radon potential map of Utah (modified from Sprinkel, 1987) showing general distribution of indoor radon concentrations determined in the survey of 1988.*

radon surveys. The cross-hatched areas primarily represent generalized outcrop patterns of radon-producing geologic formations. The boundaries are imprecise and may be revised with future, more detailed study. Areas of low radon potential may occur within cross-hatched areas. It is important to remember that this map (figure 7) only addresses some of the factors that influence the indoor radon hazard. Other factors such as radon movement through soil, permeability, building foundation condition, and indoor atmospheric pressure are not considered.

THE UTAH INDOOR RADON STUDY

Although small concentrations of radon occur virtually everywhere, parts of Utah have all of the necessary geologic conditions to identify them as potential radon-hazard areas. Elevated levels of radon in any one building, and the resultant risk posed to its occupants, are largely controlled by building construction and occupant life styles. However, indoor radon levels are consistently higher in areas where favorable geologic conditions exist (Otton, personal communication, 1988). The UBRC conducted a survey to assess indoor radon levels statewide. The information derived from this study provided the first indication of the extent of Utah's indoor radon problem, and provided the UGMS with valuable information required to examine the relation between geology and indoor radon levels.

STUDY METHODS

The indoor radon study commenced in late 1987, and 631 homes were ultimately tested. Alpha track-etch monitoring devices were provided by Terradex Corporation to volunteer homeowners. The volunteers were solicited from cities or towns within radon-hazard areas (figure 7). The homes selected to participate in the study were owner-occupied, single-family dwellings. The volunteers were instructed to place the monitors in the lowest livable area in their homes, and were asked to monitor their homes for at least twelve months. The distribution of the monitors was based on population density. Thus, the Wasatch Front (the metropolitan area from Provo to Brigham City) received about 80 percent of the monitors. Throughout the study, volunteers were regularly contacted to insure proper testing protocol. The monitoring period ended in the final quarter of 1988, and nearly every monitor was returned for analysis. The results of analyses were reported to the UBRC and the UGMS. Radon levels determined for individual homes will not be released by these agencies to the general public; survey participants received test results only for their own home. Preliminary survey results were compiled early in 1989.

Geographic distribution of the radon data was analyzed by compiling summary statistics of radon values by zip code. For some rural areas of Utah, post office box numbers made exact locations impossible to determine. Radon values between 4 to 10 pCi/l, 10.1 to 20 pCi/l, and greater than 20 pCi/l were plotted on the Potential Radon Hazard Map (figure 7) for comparison of indoor values and the mapped hazard areas. A geologic basis for clustering of high radon values was then determined by comparison of the survey data to the geologic map of Utah (Hintze, 1980), to selected regional geologic maps, and to a map of shallow ground water (Hecker and others, 1988).

STATEWIDE RESULTS

Results of the Utah indoor radon survey show a lognormal distribution with a geometric mean of 1.8 pCi/l and a maximum concentration of 68 pCi/l (table 2; appendix). Nearly 86 percent of the homes tested had concentrations less than 4 pCi/l and about 14 percent of the homes were found to have concentrations greater than 4 pCi/l (table 3). These results agree well with the earlier testing of 38 homes in Utah, 15.8 percent of which were found to have concentrations greater than 4 pCi/l (Lafavore, 1987). The 1980 census for Utah indicates about 288,000 single-family homes statewide. The survey results, therefore, show that there may be 41,100 homes with elevated indoor radon concentrations (33,400 between 4 and 10 pCi/l; 5,400 between 10 and 20; and 2,300 greater than 20). This is likely a maximum estimate of the potential hazard, because most participants were solicited from suspected radon-hazard areas delineated on the basis of geologic parameters. Within the identified hazard areas, clusters of high indoor radon values (greater than 10 pCi/l) were apparent. The clusters occurred in Monroe, Sevier County and in Wasatch Front communities in and near Provo, Sandy, and Ogden. Isolated high indoor radon values were recorded elsewhere in Utah.

Sevier County

Sevier County is principally a rural area with a small population and a low population density. Most of the residents are engaged in agriculture and related activities, but they live in towns rather than on farms. Most of the population and agricultural activity is concentrated in the central Sevier River Valley in the western part of the County. Radon survey results were received from fourteen homes located in four towns of the central Sevier River Valley: Monroe, Joseph, Richfield, and Sevier (table 3).

The three highest radon concentrations were measured in Monroe, with a maximum of 22.4 pCi/l. Sixty per cent of the homes tested in Monroe had values greater than 4 pCi/l, but this may not be statistically significant because of the small sample size. There is, however, a geologic basis for the high readings. The homes in Monroe are built on unconsolidated valley-fill material derived from calc-alkaline volcanic flow and tuff bedrock of the Marysvale volcanic field (Cunningham and others, 1983). The same geologic units commonly serve as a source for radon gas (Otton, 1988). Soils are permeable (Solomon and Klauk, 1989), and ground-water levels are greater than 10 feet (Young and Carpenter, 1965). The Sevier fault zone (a zone of normal faults) and a large thermal spring are present on the east side of town, both providing mechanisms for the transport of additional radon from deeper sources to the surface and ultimately indoors.

Test results from Joseph, Richfield, and Sevier are significantly lower (appendix). Only one home was measured in both Joseph and Sevier and no conclusion can be drawn from this small sample size. The maximum indoor radon concentration in Richfield was 5.3 pCi/l, with 30 percent of tested homes having values greater than 4 pCi/l. The large proportion of homes with values in excess of 4 pCi/l is probably influenced by the same factors as Monroe, but dilution of volcanic detritus with material derived from sedimentary bedrock in the vicinity (Steven and Morris, 1983) results in a lower maximum value. Sedimentary rocks, with exceptions discussed previously, are generally a poor source of radon. Transport of radon may also be inhibited by lower soil permeability and shallow ground water.

Wasatch Front

The Wasatch Front of north-central Utah includes most major population centers of the state and therefore the greatest number of homes potentially affected by radon. Homes are generally built on unconsolidated Quaternary basin fill, deposited from 1.6 m.y. (million years before the present) to the present, which is dominated by lake deposits that range from coarse-grained deltaic sediments to fine-grained lake-bottom sediments. Coarser grained lake sediments, with higher permeability, generally occur near the adjacent Wasatch Range and provide pathways for radon migration into buildings. The mountains contain sedimentary, igneous, and metamorphic bedrock, some of which provide a source for radon, as well as a source for radon-enriched lake deposits. The Wasatch fault zone, an active normal fault, lies at the foot of the Wasatch Range and forms the eastern boundary of the Basin-and-Range physiographic province. The fault zone provides a pathway for migration of radon gas from deeper source rocks. Ground water is shallower in valley locations and serves to inhibit radon migration.

The radon monitors were randomly distributed among Wasatch Front communities and radon concentrations were mostly less than 4 pCi/l. However, higher radon concentrations, particularly values greater than 10 pCi/l, occurred along eastern edges of Wasatch Front valleys. Areas in Sandy and Provo have apparent clusters of high indoor radon concentrations. These clusters are indicated by higher geometric means of values (2.28 pCi/l in Sandy; 2.03 pCi/l in Provo) as compared to the statewide geometric mean (1.80 pCi/l; table 2), suggesting that high arithmetic averages for the two areas (3.52 pCi/l and 3.10 pCi/l, respectively) are influenced more by many higher indoor measurements than by a single,

extremely high measurement. The latter case occurs in the Ogden area, where a high arithmetic average (3.42 pCi/l) is skewed by a single measurement of 68.2 pCi/l, but the divergence of arithmetic average and geometric mean (1.50 pCi/l) suggests that, apart from the single high measurement, most other values in Ogden were quite low (table 3).

To study this tendency of higher radon concentrations to cluster in certain locations close to the mountain front, radon concentrations in zip code areas close to the mountains were compared to those in valley zip code areas and the geology of each was examined. For Utah County, Provo (mountain front) was compared to Orem (valley); for Salt Lake County, Sandy and eastern Salt Lake Valley communities (mountain front) were compared to western Salt Lake Valley communities (valley); and for Weber County, Ogden area mountain front and valley communities were compared. Tables 2 and 3 summarize the radon concentration statistics for each area.

Utah County — In Utah County, the maximum indoor radon concentration of 13.6 pCi/l was measured in Provo (table 2). About 21 percent of homes tested in Provo have values greater than 4 pCi/l (table 3). Most of the higher values fell within zip code 84604 (figure 8) located along the mountain front (figure 9). The geology of this area consists of Lake Bonneville sediments of the Provo River delta, other nearshore lacustrine deposits, and younger alluvial deposits (Davis, 1983b; Machette, 1989). Ground water is generally greater than 10 feet deep (Anderson and others, 1986a). In addition, some of the area is underlain by the Mississippian Manning Canyon Shale, a dark marine shale enriched in uranium. The Manning Canyon Shale is also the parent material for some of the Quaternary valley-fill deposits. Most of the higher values in Utah County occur south of the mouth of Provo Canyon. This may reflect deposition of coarser grained material, derived in part from the Manning Canyon Shale by longshore currents in Lake Bonneville. Higher indoor radon levels may result from the

	UTAH	OGDEN	OREM	PROVO	SANDY	EAST SLV	WEST SLV
Sample Size	631	49	44	43	42	181	40
Average	2.73	3.42	2.12	3.10	3.52	2.24	1.68
Median	1.80	1.30	1.90	2.10	2.10	1.70	1.40
Mode	1.00	0.80	2.20	0.70	0.90	0.70	1.70
Geometric mean	1.80	1.50	1.80	2.03	2.28	1.63	1.37
Variance	18.14	96.06	1.22	9.68	20.27	3.82	1.56
Standard deviation	4.26	9.80	1.11	3.11	4.50	1.95	1.25
Standard error	0.17	1.40	0.17	0.47	0.69	0.15	0.20
Minimum	0.01	0.30	0.20	0.30	0.50	0.01	0.30
Maximum	68.20	68.20	4.60	13.60	26.20	15.70	6.50
Range	68.19	67.90	4.40	13.30	25.70	15.69	6.20
Lower quartile	1.00	0.80	1.30	0.90	1.30	1.00	0.90
Upper quartile	3.10	2.10	3.10	3.70	3.40	2.80	1.85
Interquartile range	2.10	1.30	1.80	2.80	2.10	1.80	0.95
Skewness	9.19	6.31	0.40	1.71	3.58	2.85	2.26
Standardized skewness	94.21	18.03	1.08	4.58	9.46	15.67	5.83
Kurtosis	118.98	41.94	-0.80	2.47	15.70	13.67	5.88
Standardized kurtosis	610.07	59.92	-1.09	3.31	20.76	37.55	7.58

Table 2. *Statistical analysis of indoor radon concentrations measured in the 1988 radon study conducted by the Utah Bureau of Radiation Control. Ogden area includes Ogden, North Ogden, South Ogden, Pleasant View, Washington Terrace, and Uintah; East Salt Lake Valley (East SLV) includes most cities east of the Jordan River such as Salt Lake City, South Salt Lake City, Holladay, Murray, Midvale, and Draper; West Salt Lake Valley (West SLV) includes most cities west of the Jordan River such as West Valley City, Kearns, Bennion, Taylorsville, West Jordan, South Jordan, and Riverton.*

LOCATION	TOTAL HOMES	NO. HOMES	<4 pCi/l	NO. HOMES	4<10 pCi/l	NO. HOMES	10<20 pCi/l	NO. HOMES	≥20 pCi/l
Utah	631	541	85.74%	73	11.57%	12	1.90%	5	0.79%
Ogden	49	44	89.80%	2	4.08%	2	4.08%	1	2.04%
Orem	44	43	97.73%	1	2.27%	0	0.00%	0	0.00%
Provo	43	34	79.07%	7	16.28%	2	4.65%	0	0.00%
Sandy	42	34	80.95%	6	14.29%	1	2.38%	1	2.38%
East SLV	181	159	87.85%	20	11.05%	2	1.10%	0	0.00%
West SLV	40	37	92.50%	3	7.50%	0	0.00%	0	0.00%
Sevier Co.	14	8	57.14%	3	21.43%	1	7.14%	2	14.29%

Table 3. *Distribution of indoor radon concentrations measured in the 1988 UBRC radon study. Ogden area includes Ogden, North Ogden, South Ogden, PleasantView, Washington Terrace, and Uintah; East Salt Lake Valley (East SLV) includes most cities east of the Jordan River such as Salt Lake City, South Salt Lake City, Holladay, Murray, Midvale, and Draper; West Salt Lake Valley (West SLV) includes most cities west of the Jordan River such as West Valley City, Kearns, Bennion, Taylorsville, West Jordan, South Jordan, and Riverton.*

decay of uranium in these sediments, and the relatively rapid upward migration of radon gas through permeable nearshore lacustrine and alluvial sediments and along the Wasatch fault zone.

In Orem, the maximum indoor radon concentration was 4.6 pCi/l with about 2 percent of homes tested having values greater than 4 pCi/l (table 3). Most higher values fell within zip codes 84057 (figure 10). Orem is largely located on the distal end of a Lake Bonneville delta (Davis, 1983b; Machette, 1989) which consists of finer-grained deposits and is further away from bedrock than the Provo area. Ground-water levels in the valley are generally less than 10 feet deep (Anderson and others, 1986a). Lower indoor radon concentrations in Orem are the result of the dilution of sediments derived from radon-enriched rocks with non-enriched

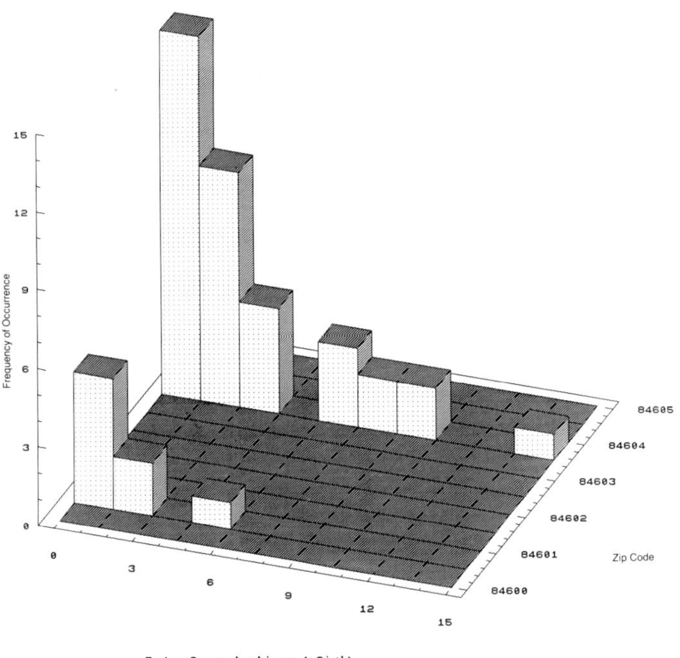

Figure 8. *Histogram of indoor radon concentrations in the Provo area. Higher concentrations are clustered in zip code areas along the mountain front. See figure 9 for a map of Provo zip code areas.*

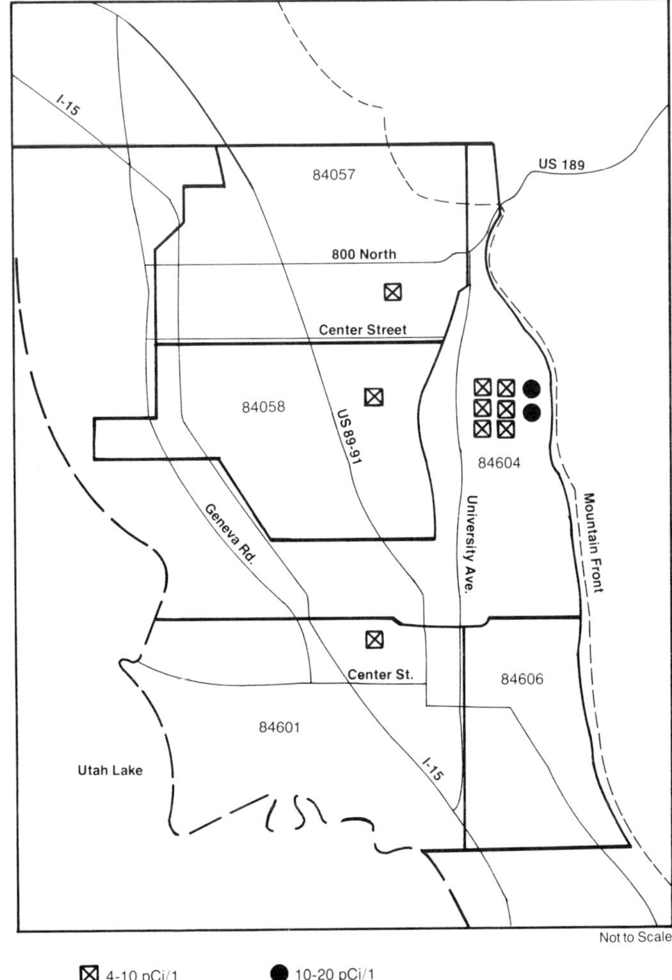

Figure 9. *Zip code map of the Provo-Orem area with distribution of elevated indoor radon concentrations. Distribution is by zip code and does not indicate specific measurement locations. Zip code boundaries are approximate.*

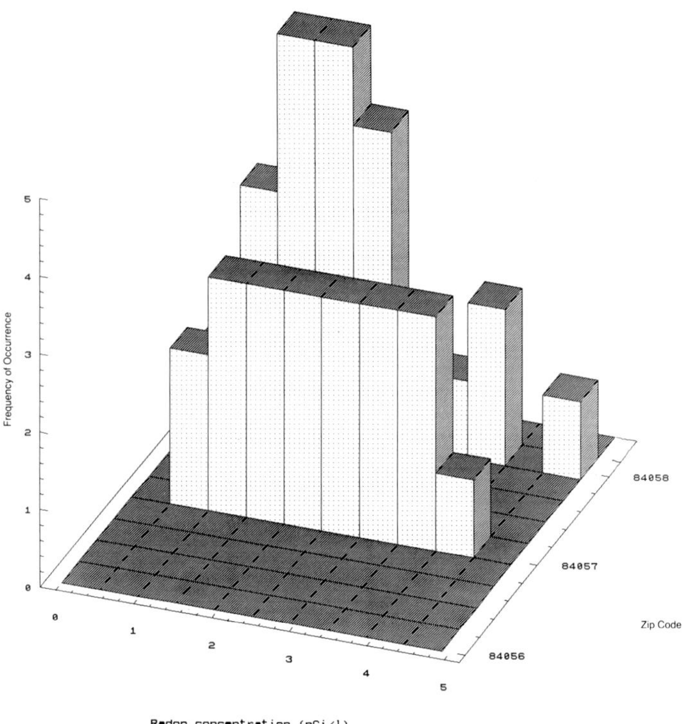

Figure 10. *Histogram of indoor radon concentrations in the Orem area. See figure 9 for a map of Orem zip code areas.*

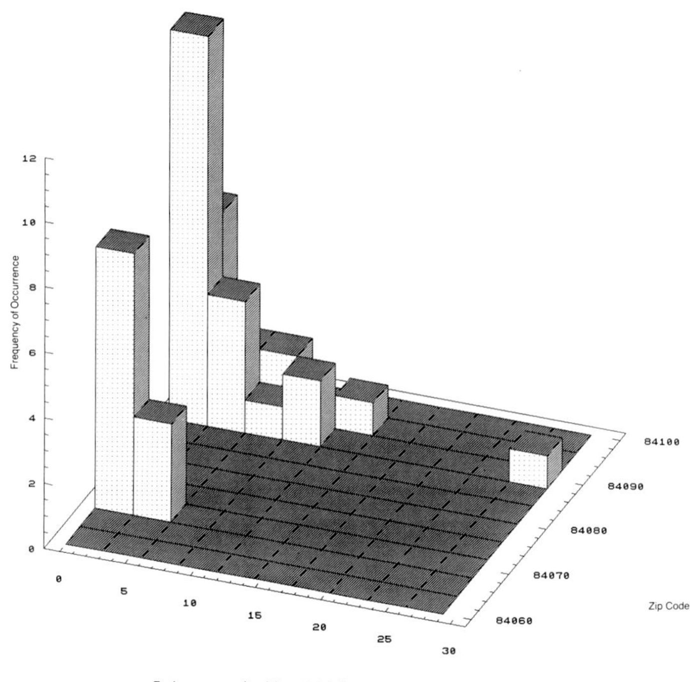

Figure 11. *Histogram of indoor radon concentrations in the Sandy area. Higher concentrations are clustered in zip code areas along the mountain front. See figure 12 for a map of Sandy zip code areas.*

material, lower permeability of finer grained lake deposits which inhibit radon migration, the presence of shallow ground water which also inhibits migration, and the absence of the Wasatch fault zone to serve as a conduit from deeper source rocks.

Salt Lake County — In Salt Lake County, the maximum indoor radon concentration of 26.2 pCi/l was measured in Sandy (table 2). About 19 percent of homes tested in Sandy have values greater than 4 pCi/l (table 3). Most of the higher values fell within zip codes 84092 and 84093 (figure 11), an area along the mountain front (figure 12). Sandy is located at the mouth of Little Cottonwood Canyon and the homes there are built on coarse-grained deltaic, glacial, and debris-flow deposits (Davis, 1983a; Personius and Scott, 1990). Near the mouth of Little Cottonwood Canyon, quartz monzonite of the Little Cottonwood stock is the source for most of the Quaternary deposits. Depth to ground water is generally greater than 10 feet (Anderson and other, 1986b). Most of the higher indoor radon values in Salt Lake County occur south of the mouth of Little Cottonwood Canyon, possibly reflecting longshore current deposition of radon-rich clastic sediments, higher permeability, and deeper ground water than in valley locations. In addition, the Wasatch fault zone trends through this area of Sandy (figure 13). The clustering of high indoor radon values in Sandy probably reflects the nearby bedrock (quartz monzonite) source of radon, relatively undiluted in coarser grained late Quaternary

units; pathways for migration of radon through coarse-grained, permeable sediment, uninhibited by shallow ground water; and location of the Wasatch fault zone as a conduit for the movement of radon gas from deeper sources to soil beneath residences.

In the western Salt Lake Valley (West Valley City, Bennion, Kearns, South Jordan, West Jordan, Riverton, and Taylorsville), the maximum indoor radon concentration was 6.5 pCi/l with 7.5 percent of the homes tested having values greater than 4 pCi/l (table 3). The higher values fell within zip codes 84065, 84118, and 84120 (figure 14) located near the center of the valley (figure 12). The western Salt Lake Valley is mostly underlain by fine-grained lake-bottom deposits of Lake Bonneville (Davis, 1983a; Personius and Scott, 1990), deposited far from the source of uranium-enriched rocks. In addition, much of the western Salt Lake Valley is located in an area of shallow ground water (Anderson and others, 1986b) which may impede the migration of radon through the soil.

In the eastern Salt Lake Valley (Salt Lake City, South Salt Lake City, Holladay, Murray, Midvale, and Draper), the maximum indoor radon concentration was 15.7 pCi/l with about 12.1 percent of the homes tested having values greater than 4 pCi/l (table 3). Most of the higher values fell within mountain front zip codes, particularly 84108, 84109, and 84124 (figure 15). In this part of the Salt Lake Valley it appears that higher concentrations of indoor

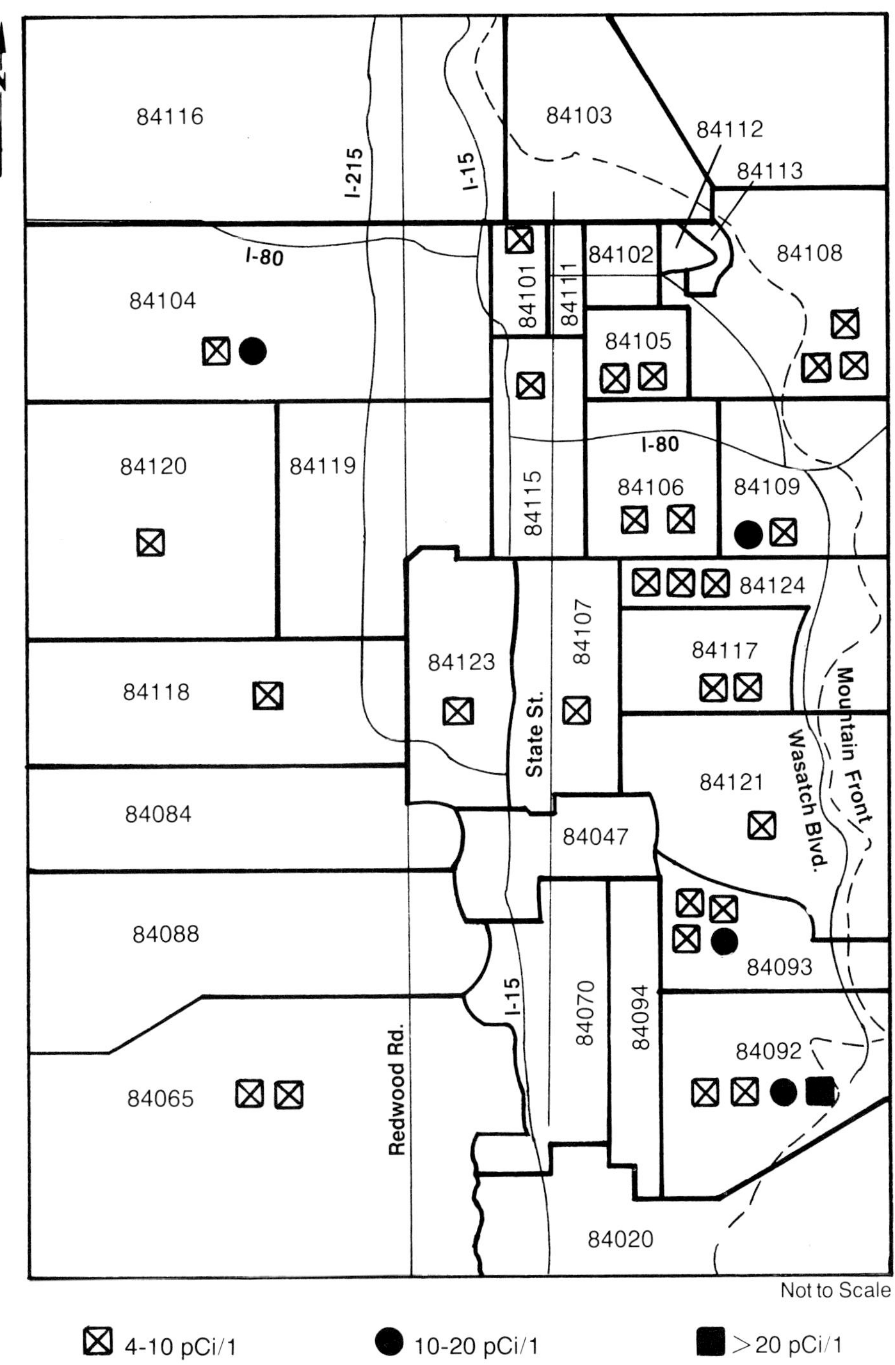

Figure 12. *Zip code map of the Salt Lake City area with distribution of elevated indoor radon concentrations. Distribution is by zip code and does not indicate specific measurement locations. Zip code boundaries are approximate.*

radon are located near the mountains and consistently lower values dominate valley locations (figure 12). Similar to observations in the southern part of the Salt Lake Valley, the mountain front zip codes consist of mostly coarser grained deltaic deposits and the valley zip codes are dominated by finer grained lake-bottom deposits (Davis, 1983a; Personius and Scott, 1990). Ground water is less than 10 feet deep in valley locations (Anderson and others, 1986b),

and the Wasatch fault zone ruptures surficial sediments in mountain front locations.

Weber County — In Weber County, the maximum indoor radon concentration of 68.2 pCi/1 was measured in Uintah, the highest recorded value in Utah to date (table 2). Uintah is in the Ogden metropolitan area (Ogden, North Ogden, South Ogden, Pleasant View, Washington Terrace, and Uintah), which has about 10.2

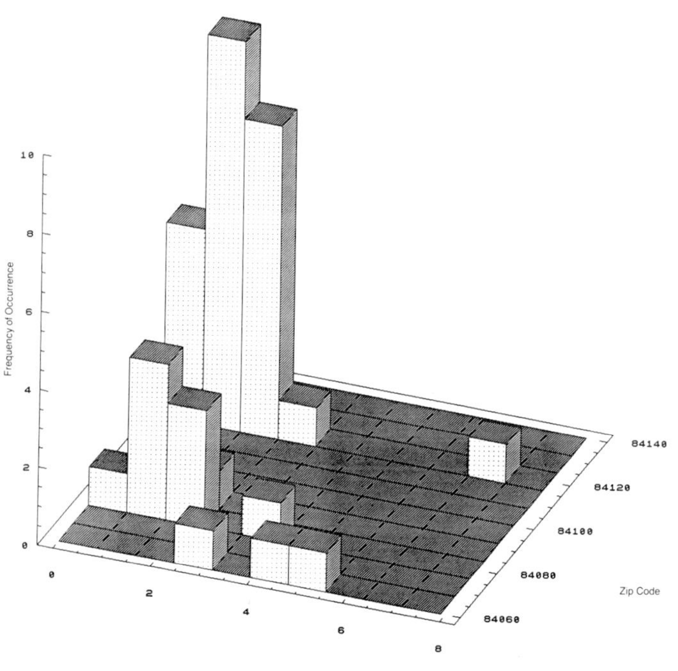

Figure 13. *The mouth of Little Cottonwood Canyon. Quartz monzonite of the Little Cottonwood Canyon stock forms hills at the mouth of the canyon and serves as a source of radon gas. Deltaic sediments on the east bench were derived from the stock and were distributed by longshore currents along the shore of Lake Bonneville. Homes are built on these sediments, as well as on glacial and debris-flow deposits. The Wasatch fault zone (shown by arrows) separates the mountains from the valley and serves as a conduit for radon gas to travel from depth to the surface.*

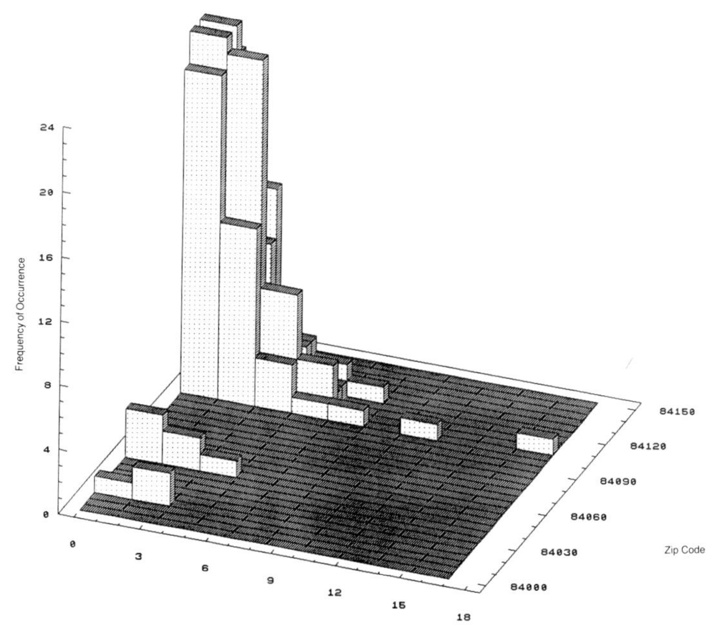

Figure 14. *Histogram of indoor radon concentrations in the western Salt Lake Valley. See figure 12 for a map of western Salt Lake Valley zip code areas.*

Figure 15. *Histogram of indoor radon concentrations in the eastern Salt Lake Valley. Higher concentrations are clustered in zip code areas along the mountain front. See figure 12 for a map of eastern Salt Lake Valley zip code areas.*

percent of tested homes with values greater than 4 pCi/l (table 3). As in the Salt Lake metropolitan and Provo areas, higher concentrations of indoor radon in the Ogden area are located along the mountain front, in zip codes 84403 and 84405 (figure 16), and lower values dominate valley locations (figure 17). However, the incidence of high indoor radon concentrations in the Ogden area is infrequent when compared to the other areas. The Ogden area was thought to have an equal or greater probability of elevated indoor radon because mountains adjacent to the southern part of the area are dominated by the Precambrian Farmington Canyon Complex, which served as a sediment source for Quaternary sedimentary deposits (Davis, 1985; Nelson and Personius, in press). These Precambrian metamorphic rocks consist of argillite, gneiss, and schist and are thought to be excellent sources of radon. Ogden Canyon, though, also drains highlands which are predominantly underlain by limestone and quartzite, two rock types not normally uranium-enriched. Moreover, deltaic sediments near the mouth of Ogden Canyon contain much fine-grained material and are relatively impermeable when compared to deltaic sediments near Provo and Sandy; this inhibits the migration of soil gas. Survey results in Ogden may reflect the need for a more precise determination of the relation between geology and indoor radon levels.

Levels in excess of 10 pCi/l were also measured in Huntsville and Roy, Weber County. These are isolated occurrences with no statistical significance and a geologic basis for these levels was not investigated.

Other Counties

Indoor radon levels in excess of 10 pCi/l were measured in other communities throughout Utah. These communities include Beaver, Beaver County; Park Valley, Box Elder County; Laketown, Rich County; and New Harmony, Washington County. As with Huntsville and Roy, Weber County, these measurements are iso-

lated and no geologic basis for them has been investigated. A complete list of all survey measurements is found in the appendix to this report.

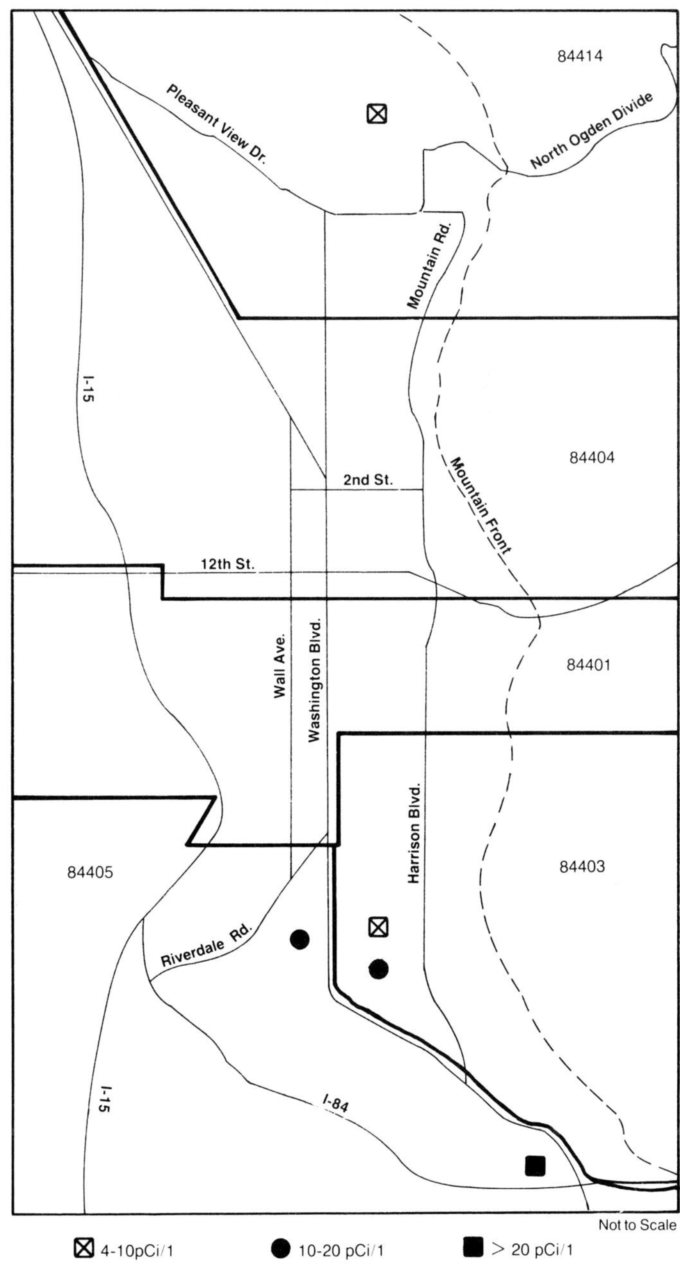

Figure 17. *Zip code map of the Ogden area with distribution of elevated indoor radon concentrations. Distribution is by zip code and does not indicate specific measurement locations. Zip code boundaries are approximate.*

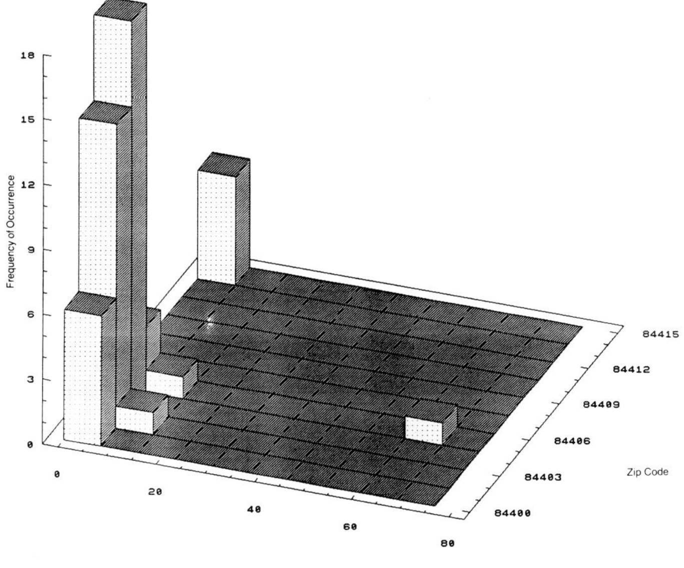

Figure 16. *Histogram of indoor radon concentrations in the Ogden area. See figure 17 for a map of Ogden zip code areas.*

CURRENT RADON STUDIES

The UGMS, in cooperation with the University of Utah Research Institute (UURI), conducted an investigation on Antelope Island in 1989 to add to the understanding of the geologic factors that influence radon occurrence, emanation, and migration. Antelope Island was selected because detailed geologic mapping (Doelling and others, 1988) shows a variety of structurally complex metamorphic, igneous, and sedimentary rocks. Many of the metamorphic and igneous rocks could serve as a source of radon gas. The study consisted of the measurement along several traverses across the island of radon in soil gas with a radon emanometer, a statistical analysis of the collected data, and correlation with potential source rocks, soil types, and hydrologic parameters. Final analysis is not complete, but the study will ultimately lead to a refinement of test methodology required for future site characterization studies elsewhere. The geology on the island is similar in some respects to that of Davis and Weber Counties, and this study will aid in the greater understanding of the potential radon hazard of this part of the Wasatch Front urban corridor.

Detailed studies will also be conducted in the Wasatch Front region with funding obtained through the EPA State Indoor Radon Grant Program The UBRC has solicited the participation of 400 volunteers in the Sandy and Provo areas to monitor their homes for indoor radon during a year-long study. Concurrently, the UGMS will investigate geologic factors that influence indoor radon concentrations by the measurement, on a grid pattern, of radioactive soil material with a portable gamma-ray spectrometer, radon gas with a radon emanometer, and soil moisture and density with a portable moisture-density meter. These geologic studies will define the distribution of the radioactive source in soil, the distribution of radon gas derived from the source, and the effect of soil moisture and permeability on the migration of radon. These studies should help explain why the Provo and Sandy areas were "hot spots" in the statewide survey of 1988, test geologic models of radionuclide distribution in sediments, and identify relevant geologic factors that may be used as predictive tools to aid in identifying areas where radon levels may be high and mitigation necessary.

SUMMARY

Radon is an environmental concern throughout the country because of its suspected link to lung cancer. Radon is an odorless, tasteless, and colorless radioactive gas that occurs in nearly all rocks and soils. It is found in most buildings in small concentrations that do not constitute a health threat. However, scientists have recently discovered that geologic conditions can influence the likelihood of having elevated indoor radon levels.

A statewide, year-long study documented areas of elevated indoor radon concentrations in Utah. The distribution of concentrations was lognormal, and nearly 86 percent of the homes tested had indoor radon concentrations less than 4 pCi/l. Anomalous areas of elevated indoor radon concentrations were found in the Wasatch Front communities of Sandy and Provo. In these communities, homes near the mountain front are more likely to have elevated indoor radon levels than homes in valley areas. Sandy and Provo appear to have geologic factors that control radon distribu-

tion. Both communities are located on high permeability, proximal deltaic deposits of Lake Bonneville, have a known bedrock radon source, have ground-water depths greater than 10 feet, and are near the Wasatch fault zone. However, further study is needed to fully understand both the geologic and non-geologic factors that control these anomalies. Homes in the Monroe area of Sevier County also have elevated indoor radon levels, and geology appears to control clustering there, too. Monroe is located on permeable alluvial sediments derived from a known bedrock radon source, has ground-water depths greater than 10 feet, and is near the Sevier fault zone and a large thermal spring. Immediate efforts to study the relationship between geology and indoor radon levels will be concentrated along the heavily populated Wasatch Front.

Because of the complex relationships between geologic and non-geologic factors that control radon levels, predicting radon concentrations from building to building is difficult even in areas with a high geologic potential for radon production. The current understanding of radon behavior prohibits extrapolating radon values over any distance. But with additional indoor radon surveys and geologic characterization of sites, discovering critical combinations of components will lead to an easier and more reliable method of radon assessment. It is important to determine the critical factors that contribute to the potential radon hazard for areas prior to construction so that mitigation techniques can be incorporated into building design.

ACKNOWLEDGMENTS

We are indebted to the following for taking time to review an earlier version of this manuscript. Their efforts are appreciated and significantly improved this paper: Dane Finerfrock, UBRC; Dennis L. Nielson, UURI; Miriam H. Bugden, UGMS; John S. Hand, Cypress Minerals (formerly with UGMS); and Genevieve Atwood, formerly with UGMS. We would like to especially thank James K. Otton of the U.S. Geological Survey for taking the time to review the earlier version of this manuscript. His guidance and suggestions were particularly helpful. Gary E. Christenson and William R. Lund of the UGMS reviewed this report and provided valuable comments.

REFERENCES

Adams, J.A.S., and Lowder, W.M., 1964, editors. The natural radiation environment: Chicago, Illinois, University of Chicago Press, 1069 p.

Adams, J.A.S., Lowder, W.M., and Gesell, T.F., 1972, editors, The natural radiation environment II: Springfield, Virginia, National Technical Information Service, United States Energy Research and Development Agency Report CONF-720805.

Anderson, L.R., Keaton, J.R., and Bischoff, J.E., 1986a, Liquefaction potential map for Utah County, Utah: Prepared by Utah State University Department of Civil and Environmental Engineering and Dames and Moore Consulting Engineers for U.S. Geological Survey, Contract No. 14-08-0001-21359, 46 p.

Anderson, L.R., Keaton, J.R., Spitzley, J.E., and Allen, A.C., 1986b, Liquefaction potential map for Salt Lake County, Utah: prepared by Utah State University Department of Civil and Environmental Engineering and Dames and Moore Consulting Engineers for U.S. Geological Survey, Contract No. 14-08-000119910, 48 p.

Austin, L.H., 1988, Problems and management alternatives related to the rising level of the Great Salt Lake, *in* Proceedings 6th International Water Resources Association World Congress on Water Resources: International Water Resources Association, Ottawa, Canada, May 29-June 3, 1988.

Barretto, P.M.C., 1975, ^{222}Rn emanation characteristics of rocks and minerals, *in* Proceedings of a Panel of Radon in Uranium Mining: Vienna, Austria, International Atomic Energy Agency, STI/PUB/391, p. 129-148, Discussion, p. 148-150.

Chenoweth, W.L., 1975, Uranium deposits of the Canyonlands area, *in* Fassett, J.E., editor, Canyonlands Country: Four Corners Geologic Society 8th Annual Field Conference Guidebook, p. 253-260.

Clements, W.E., and Wilkening, M.H., 1974, Atmospheric pressure effects on ^{222}Rn transport across earth-air interface: Journal of Geophysical Research, v. 79, no. 33, p. 5025-5029.

Cross, F.T., Harley, N.H., and Hofman, W., 1985, Health effects and risks from ^{222}Rn in drinking water: Health Physics, vol. 48, no. 5, p. 649.

Cunningham, C.G., Steven, T.A., Rowley, P.D., Glassgold, L.B., and Anderson, J.J., 1983, Geologic map of the Tushar Mountains and adjoining areas, Marysvale volcanic field, Utah: U.S. Geological Survey Miscellaneous Investigations Map I-1430-A, scale 1:50:000.

Davis, F.D., 1983a, Geologic map of the central Wasatch Front, Utah: Utah Geological and Mineral Survey Map 54-A, scale 1:100,000.

---1983b, Geologic map of the southern Wasatch Front, Utah: Utah Geological and Mineral Survey Map 55-A, scale 1:100,000.

---1985, Geologic map of the northern Wasatch Front, Utah: Utah Geological and Mineral Survey Map 53-A, scale 1:100,000.

Doelling, H.H., 1969, Mineral resources, San Juan County, Utah, and adjacent areas, Part II — Uranium and other metals in sedimentary host rocks: Utah Geological and Mineral Survey Special Studies 24, 64 p.

Doelling, H.H., Willis, G.C., Jensen, M.E., Hecker, Suzanne, Case, W.F., and Hand, J.S., 1988, Geologic map of Antelope Island, Davis County, Utah: Utah Geological and Mineral Survey Open-File Report 144, 99 p., scale 1:24,000.

Durrance, E.M., 1986, Radioactivity in geology, principles and applications: New York City, New York, John Wiley and Sons, 441 p.

Duval, J.S., Jones, W.J., Riggle, F.R., and Pitkin, J.A., 1989, Equivalent uranium map of the conterminous United States: U.S. Geological Survey Open-File Report 89-478.

Environmental Protection Agency, 1986a, A citizens guide to radon, what it is and what to do about it: Environmental Protection Agency and Center for Disease Control, OPA-86-004, 13 p.

---1986b, Interim radon and radon decay product measurement protocols: Environmental Protection Agency, Office of Radiation Programs, EPA 520/1-8-04.

---1987, Radon reference manual: Environment Protection Agency, Office of Radiation Programs. EPA 520/1-87-20.

Fleischer, R.L., Mogro-Compero, Antonio, and Turner, L.G., 1982, Radon levels in homes in the northeastern United States -Energy-efficient homes, *in* Vohra, K.G., Mishra, U.C., Pillai, K.C., and Sadasivan, S., editors, Natural Radiation Environment: New Delhi, India, Wiley Eastern Ltd., p. 497-502.

Gesell, T.F., and Lowder, W.M., 1980, editors, Natural radiation environment III: Springfield, Virginia, National Technical Information Services, United States Department of Energy Symposium Series 51 CONF-780422, 1739 p.

Gundersen, L.C.S., 1990, Grain size and emanation as controlling factors in soil radon, *in* U.S. Environmental Protection Agency: The 1990 International Symposium on Radon and Radon Reduction Technology, Atlanta, Georgia Preprints: EPA/600/9-90/005a, vol. III, p. C-VI-4.

Gurgel, K.D., editor, 1983, Energy resources map of Utah: Utah Geological and Mineral Survey Map 68, scale 1:500,000.

Hecker, Suzanne, Harty, K.M., and Christenson, G.E., 1988, Shallow ground water and related hazards in Utah: Utah Geological and Mineral Survey Map 110, scale 1:750,000.

Heier, K.S., and Carter, J.L., 1964, Uranium, thorium, and potassium contents in basic rocks and their bearing on the nature of the upper mantle, *in* Adams, J.A.S., and Lowder, W.M., editors. The Natural Radiation Environment: Chicago, Illinois, University Chicago Press, p. 63-86.

Hintze, L.F., 1967, Uranium districts of southeastern Utah: Utah Geological Society Guidebook to the Geology of Utah No. 21, 194 p.

Hintze, L.F., 1980, Geological map of Utah: Utah Geological and Mineral Survey Map A-1, scale 1:500,000.

Horton, T.R., 1985, Nationwide occurrence of radon and other natural radioactivity in public water supplies: Environmental Protection Agency, 520/5-85-008, 208 p.

International Atomic Energy Agency, 1976, Exploration for uranium ore deposits: Vienna, Austria, International Atomic Energy Agency, Symposium Proceedings, 807 p.

Jacobi, W., and Eisfeld, K., 1982, Internal dosimetry of ^{222}Rn, ^{220}Rn, and their short-lived daughters, *in* Vohra, K.G., Mishra, U.C., Pillai, K.C., and Sadasivan, S., editors, Natural Radiation Environment: New Delhi, India, Wiley Eastern Ltd, p. 131-143.

King, D., 1986, Gas geochemistry applied to earthquake research --an overview: Journal of Geophysical Research, v. 91, no. B12, p. 12269-12281.

Kramer, H.W., Schroeder, G.L., and Evans, R.D., 1964, Measurements of the effects of atmosphere variables on ^{222}Rn flux and soil gas concentrations, *in* Adams, J.A.S., and Lowder, W.M., editors, The Natural Radiation Environment: Chicago, Illinois, University of Chicago Press, 1069 p.

Kunz, C., Laymon, C.A., and Parker, C., 1989, Gravelly soils and indoor radon, *in* U.S. Environmental Protection Agency: The 1988 Symposium on Radon and Radon Reduction Technology, Denver, Colorado — Preprints: EPA/600/9-89/006a, vol. I, p. V-6.

Lafavore, Michael, 1987, Radon — the invisible threat: Emmaus, Pennsylvania, Rodale Press, 256 p.

Machette, M.N., 1989, Preliminary surficial geologic map of the Wasatch fault zone, eastern part of Utah Valley, Utah County and parts of Salt Lake and Juab Counties, Utah: U.S. Geological Survey Miscellaneous Field Studies Map MF-2109, scale 1:50,000.

National Research Council, 1988, Health risks of radon and other internally deposited alpha-emitters: Washington, D.C., Committee on the Biological Effects of Ionizing Radiation IV, National Academy of Sciences, 602 p.

National Council on Radiation Protection and Measurements, 1975, Natural background radiation in the United States: Bethesda, Maryland, National Council on Radiation Protection and Measurements Report 45.

---1984a, Exposures from the uranium series with emphasis on radon and its daughters: Bethesda, Maryland, National Council on Radiation Protection and Measurements Report 77, 132p.

---1984b, Evaluation of occupational and environmental exposures to radon and radon daughters in the United States: Bethesda, Maryland, National Council on Radiation Protection and Measurements Report 78, 204 p.

Nelson, A.R., and Personius, S.F., in press, Preliminary surficial geologic map of the Weber segment and adjacent parts of the Brigham City and Salt Lake City segments, Wasatch fault zone, Weber and Box Elder Counties, Utah: U.S. Geological Survey Miscellaneous Field Studies Map, scale 1:50,000.

Nero, A.V., 1986, The indoor radon story: Technology Review, v. 89, no. 1, p. 28-31, 36-40.

Nero, A.V., Boegel, M.L., Hollowell, C.D., Ingersoll, S.G., Nazaroff, W.W., and Revzan, K.L., 1982, Radon and its daughters in energy-efficient buildings, *in* Vohra, K.G., Mishra, U.C., Pillai, K.C., Sadasivan, S., editors, Natural Radiation Environment: New Delhi, India, Wiley Eastern Ltd., p. 473-480.

Nero, A.V., Schwehr, M.B., Nazaroff, W.W., and Revzan, K.L., 1986, Distribution of airborne ^{222}Rn concentrations in U.S. homes: Science, vol. 234, p. 992-997.

Otton, J.K., 1988, Potential for indoor radon hazards -a first geologic estimate, *in* Makofske, W.H., and Edelstein, M.R., editors, Radon and the Environment: Park Ridge, New Jersey, Noyes Publication, 456 p.

Otton, J.K., and Duval, J.S., 1990, Geologic controls on indoor radon in the Pacific northwest, *in* U.S. Environmental Protection Agency: The 1990 International Symposium on Radon and Radon Reduction Technology, Atlanta, Georgia — Preprints: EPA/600/9-90/005a, vol. III, p. VI-5.

Personius, S.F., 1988, Preliminary surficial geologic map of the Brigham City segment and adjacent parts of the Weber and Collinston segments, Wasatch fault zone, Box Elder and Weber Counties, Utah: U.S. Geological Survey Miscellaneous Field Studies Map MF-2042, scale 1:50,000.

Personius, S.F., and Scott, W.E., 1990, Preliminary surficial geologic map of the Salt Lake City segment and parts of adjacent segments of the Wasatch fault zone, Davis, Salt Lake, and Utah Counties, Utah: U.S. Geological Survey Miscellaneous Field Studies Map MF-2114, scale 1:50,000.

Phair, George, and Gottfried, David, 1964, The Colorado front range, Colorado, USA, as a uranium and thorium province, *in* Adams, J.A.S., and Lowder, W.M., editors, The Natural Radiation Environment: Chicago, Illinois, University of Chicago Press, p. 7-38.

Prichard, H.M., 1987, The transfer of radon from domestic water to indoor air: Journal of American Water Works Association, vol. 159.

Richardson, K.A., 1964, Thorium, uranium, and potassium in the Conway Granite, New Hampshire, USA, *in* Adams, J.A.S., and Lowder, W.M., editors, The Natural Radiation Environment: Chicago, Illinois, University of Chicago Press, p. 38-50.

Rogers, A.S., 1956, Application of radon concentrations to ground-water studies near Salt Lake City and Ogden, Utah: Geological Society of America Bulletin, v. 67, no. 12, pt. 2, p. 1781.

---1958, Physical behavior and geologic control of radon in mountain streams: U.S. Geological Survey Bulletin 1052-E, p. 187-211.

Rogers, J.J.W., 1964, Statistical test of the homogeneity of the radioactive components of granites, *in* Adams, J.A.S., and Lowder, W.M., editors, The Natural Radiation Environment: Chicago, Illinois, University of Chicago Press, p. 51-62.

Ronca-Battista, Melinda, 1988, Interim indoor radon and radon decay product measurement protocols, *in* Makofske, W.J., and Edelstein, M.R., editors, Radon and the Environment: Park Ridge, New Jersey, Noyes Publications, p. 142-146.

Samet, J.M., 1989, Radon exposure and lung cancer risk, *in* U.S. Environmental Protection Agency: The 1988 Symposium on Radon and Radon Reduction Technology, Denver, Colorado — Preprints: EPA/600/9-89/006a, vol. I, p. II-1.

Schery, S.D., and Gaeddert, D.H., 1982, Measurements of the effect of cyclic atmospheric pressure variation on the flux of 222Rn from the soil: Geophysical Research Letters, v. 9, no. 8, p. 835-838.

Schery, S.D., and Siegel, D., 1986, The role of channels in transport of radon from soils: Journal of Geophysical Research, v. 91, No. B-12, p. 12366-12374.

Schmidt, Anita, Puskin, J.S., Nelson, Christopher, and Nelson, Neal, 1990, Estimate of annual radon-induced lung cancer deaths — EPA's approach, *in* U.S. Environmental Protection Agency: The 1990 International Symposium on Radon and Radon Reduction Technology, Atlanta, Georgia — Preprints: EPA/600/9-90/005a, vol. I, p. II-3.

Schoenberg, Janet, Klotz, Judith, Wilcox, Homer, and Nicholls, Gerald, 1990, Radon and lung cancer among New Jersey women, *in* U.S. Environmental Protection Agency: The 1990 International Symposium on Radon and Radon Reduction Technology, Atlanta, Georgia Preprints: EPA/600/9-90/005a, vol. I, p. II-2.

Scott, W.E., and Shroba, R.R., 1985, Surficial geological map of an area along the Wasatch fault zone in the Salt Lake Valley, Utah: U.S. Geological Survey Open-file Report 85-448, 18 p., scale 1:24,000.

Sextro, Richard, 1988, Radon in dwellings, *in* Makofske, W.J., and Edelstein, M.R., editors, Radon and the Environment: Park Ridge, New Jersey, Noyes Publications, p. 71-82.

Sextro, R.G., Nazaroff, W.W., and Turk, B.H, 1989, Soil permeability and radon concentration measurements and a technique for predicting the radon source potential of soil, *in* U.S. Environmental Protection Agency: The 1988 Symposium on Radon and Radon Reduction Technology, Denver, Colorado — Preprints: EPA/600/9-89/006a, vol. I, p. V-5.

Silver, L.T., Williams, S., and Woodhead, J.A., 1980, Uranium in granites from southwestern United States -actinide parent-daughter system, sites, and mobilization, first year report: Pasadena, California, Division of Geological and Planetary Sciences, California Institute of Technology, 380 p.

Smith, R.C, III, Reilly, M.A., Rose, A.W., Barnes, J.H., and Berkheiser, S.W., Jr., 1987, Radon — a profound case: Pennsylvania Geology, v. 18, no. 2, p. 3-7.

Solomon, B.J., and Klauk, R.H., 1989, Regional assessment of geologic conditions for sanitary landfills in Sevier County, Utah: Utah Geological and Mineral Survey Special Studies 71, 15 p.

Sprinkel, D.A., 1987 (revised 1988), The potential radon hazard map, Utah: Utah Geological and Mineral Survey Open-File Report 108, 4 p., scale 1:1,000,000.

Sprinkel, D.A., 1988, Assessing the radon hazard in Utah: Utah Geological and Mineral Survey Survey Notes, v. 22, p. 3-13.

Sprinkel, D.A., Hand, J.S., Solomon, B.J., and Finerfrock, Dane, 1989, Correlation of geology and indoor radon levels in Utah: Geological Society of America Abstracts with Programs, v. 21, no. 6.

Steven, T.A., and Morris, H.T., 1983, Geology of the Richfield 1°x2° Quadrangle, west-central Utah: U.S. Geological Survey Open-File Report 83-583, 22 p., scale 1:250,000.

Steven, T.A., and Morris, H.T., 1984, Mineral resource potential of the Richfield 1°x2° Quadrangle, west-central Utah: U.S. Geological Survey Open-File Report 84-521, 53 p., scale 1:250,000.

Swedjemark, G.A., 1980, Radon in dwellings in Sweden, *in* Gesell, T.F., and Lowder, W.M., editors, Natural Radiation Environment III, vol. II: Springfield, Virginia, National Technical Information Services, United States Department of Energy Symposium Series 51 CONF-780422, p 1237-1259.

Tanner, A.B., 1964, Physical and chemical control on distribution of ^{226}Ra and ^{222}Rn in ground water near Great Salt Lake, Utah, *in* Adams, J.A.S., and Lowder, W.M., editors, The Natural Radiation Environment: Chicago, Illinois, University of Chicago Press, p. 253-278.

---1980, radon migration in the ground - supplementary review, *in* Gesell, T.F., and Lowder, W.M., editors, Natural Radiation Environment III, vol. I: Springfield, Virginia, National Technical Information Services, United States Department of Energy Symposium Series 51, CONF-780422, p. 5-56.

---1986, Indoor radon and its sources in the ground: U.S. Geological Survey Open-File Report 86-222, 5p.

---1990, the role of diffusion in radon entry into houses, *in* U.S. Environmental Protection Agency: The 1990 International Symposium on Radon and Radon Reduction Technology, Atlanta, Georgia Preprints: EPA/600/9-90/005a. vol.III,p. V-2.

Teng, J., and Lang, F.S., 1986, Research on ground-water radon as a fluid phone precursor to earthquakes: Journal of Geophysical Research, v. 91, no. B-12, p. 12305-12313.

Thomas, D.M., and Cuff, K.E., 1986, The association between ground gas radon variations and geologic activity in Hawaii: Journal of Geophysical Research, v. 91, no. B-12, p. 12186-12198.

Vitz, Ed, 1989, Preliminary results of a nationwide waterborne radon survey, *in* U.S. Environmental Protection Agency: The 1988 Symposium on Radon and Radon Reduction Technology, Denver, Colorado Preprints: EPA/600/9-89/006a, vol. II, p. VIII-1.

Vohra, K.G., Mishra, U.C., Pillai, K.C., and Sadasivan, S., editors, 1982. The Natural Radiation Environment: New Delhi, India, Wiley Eastern Ltd, 691 p.

Woolf, Jim, 1987, Levels of radon gas high in 13 of 31 homes surveyed in valley: Salt Lake Tribune, v. 234, no. 85, p. B1.

Young, R.A., and Carpenter, C.H., 1965, Ground-water conditions and storage in the central Sevier Valley, Utah: U.S. Geological Survey Water-Supply Paper 1787, 95 p.

Appendix

Results, 1988 Utah Bureau of Radiation Control
indoor radon survey

County	City	Zipcode	pCi/l	County	City	Zipcode	pCi/l
BEAVER	BEAVER	84713	10.5	DAVIS	FRUIT HEIGHTS	84037	0.7
BEAVER	MINERSVILLE	84752	2.9	DAVIS	FRUIT HEIGHTS	84037	0.7
BOX ELDER	BRIGHAM CITY	84302	1.1	DAVIS	FRUIT HEIGHTS	84037	1.1
BOX ELDER	BRIGHAM CITY	84302	2.0	DAVIS	KAYSVILLE	84037	1.0
BOX ELDER	BRIGHAM CITY	84302	1.7	DAVIS	KAYSVILLE	84037	0.6
BOX ELDER	BRIGHAM CITY	84302	5.8	DAVIS	LAYTON	84040	1.1
BOX ELDER	BRIGHAM CITY	84302	1.3	DAVIS	LAYTON	84040	1.2
BOX ELDER	BRIGHAM CITY	84302	2.4	DAVIS	LAYTON	84040	2.3
BOX ELDER	BRIGHAM CITY	84302	3.7	DAVIS	LAYTON	84041	0.4
BOX ELDER	BRIGHAM CITY	84302	3.7	DAVIS	LAYTON	84041	1.1
BOX ELDER	CORINNE	84307	3.8	DAVIS	LAYTON	84041	1.4
BOX ELDER	CORINNE	84307	1.9	DAVIS	LAYTON	84041	1.0
BOX ELDER	FIELDING	84311	1.8	DAVIS	LAYTON	84041	1.2
BOX ELDER	GARLAND	84312	3.0	DAVIS	SUNSET	84015	2.6
BOX ELDER	GROUSE CREEK	84313	7.9	DAVIS	SYRACUSE	84075	0.9
BOX ELDER	HONEYVILLE	84314	0.9	DAVIS	W. BOUNTIFUL	84087	0.2
BOX ELDER	PARK VALLEY	84329	1.8	DAVIS	WOODS CROSS	84087	1.8
BOX ELDER	PARK VALLEY	84329	52.0	DUCHESNE	BLUEBELL	84007	2.2
CACHE	BENSON	84335	1.3	DUCHESNE	BLUEBELL	84007	0.6
CACHE	HYDE PARK	84318	0.5	DUCHESNE	BLUEBELL	84007	1.7
CACHE	HYDE PARK	84318	4.1	DUCHESNE	BLUEBELL	84007	0.3
CACHE	HYDE PARK	84318	1.9	DUCHESNE	BLUEBELL	84007	0.6
CACHE	LEWISTON	84320	1.0	DUCHESNE	DUCHESNE	84021	1.1
CACHE	LOGAN	84321	4.2	DUCHESNE	DUCHESNE	84021	0.3
CACHE	LOGAN	84321	0.7	DUCHESNE	DUCHESNE	84021	5.7
CACHE	LOGAN	84321	3.6	DUCHESNE	DUCHESNE	84021	2.7
CACHE	LOGAN	84321	5.9	DUCHESNE	FRUITLAND	84007	1.9
CACHE	LOGAN	84321	1.4	DUCHESNE	MYTON	84052	0.6
CACHE	LOGAN	84321	2.3	DUCHESNE	NEOLA	84052	1.0
CACHE	LOGAN	84321	2.5	DUCHESNE	ROOSEVELT	84066	3.0
CACHE	LOGAN	84322	7.1	DUCHESNE	ROOSEVELT	84066	2.9
CACHE	MILLVILLE	84326	3.4	GARFIELD	ESCALANTE	84726	6.4
CACHE	PROVIDENCE	84332	2.2	GARFIELD	PANQUITCH	84759	3.2
CACHE	RIVER HEIGHTS	84321	0.8	GRAND	MOAB	84532	0.7
CACHE	SMITHFIELD	84335	1.2	GRAND	MOAB	84532	5.6
CARBON	PRICE	84501	0.4	IRON	CEDAR CITY	84720	1.8
DAVIS	BOUNTIFUL	84010	4.3	IRON	CEDAR CITY	84720	1.1
DAVIS	BOUNTIFUL	84010	1.1	IRON	CEDAR CITY	84720	2.1
DAVIS	BOUNTIFUL	84010	1.2	IRON	CEDAR CITY	84720	0.6
DAVIS	BOUNTIFUL	84010	2.9	IRON	CEDAR CITY	84720	1.5
DAVIS	BOUNTIFUL	84010	1.6	IRON	PARAGONAH	84760	3.8
DAVIS	BOUNTIFUL	84010	1.0	KANE	KANAB	84741	0.5
DAVIS	BOUNTIFUL	84010	1.4	KANE	ORDERVILLE	84758	1.9
DAVIS	BOUNTIFUL	84010	1.5	MILLARD	DELTA	84624	0.3
DAVIS	BOUNTIFUL	84010	1.5	MILLARD	OAK CITY	84649	1.0
DAVIS	BOUNTIFUL	84010	2.7	MORGAN	MORGAN	84050	5.7
DAVIS	BOUNTIFUL	84010	1.1	MORGAN	MORGAN	84050	2.2
DAVIS	CENTERVILLE	84014	2.9	MORGAN	MTN GREEN	84050	3.3
DAVIS	CENTERVILLE	84014	3.4	PIUTE	CIRCLEVILLE	84723	2.1
DAVIS	CENTERVILLE	84014	1.4	RICH	GARDEN CITY	84028	2.3
DAVIS	CENTERVILLE	84014	3.1	RICH	GARDEN CITY	84028	1.6
DAVIS	CLEARFIELD	84015	1.0	RICH	LAKETOWN	84038	6.6
DAVIS	CLINTON	84015	0.2	RICH	LAKETOWN	84038	12.1
DAVIS	CLINTON	84015	0.7	RICH	RANDOLPH	84064	1.9
DAVIS	FARMINGTON	84025	2.8	RICH	RANDOLPH	84064	2.5
DAVIS	FARMINGTON	84025	0.7	RICH	RANDOLPH	84064	1.6
DAVIS	FRUIT HEIGHTS	84037	0.3	RICH	WOODRUFF	84086	2.2

County	City	Zipcode	pCi/l
RICH	WOODRUFF	84086	2.2
RICH	WOODRUFF	84086	1.8
SALT LAKE	BENNION	84118	1.7
SALT LAKE	BRIGHTON	84121	1.2
SALT LAKE	CANYON	84121	6.8
SALT LAKE	DRAPER	84020	2.7
SALT LAKE	DRAPER	84020	3.2
SALT LAKE	DRAPER	84020	0.8
SALT LAKE	HOLLADAY	84117	1.0
SALT LAKE	KEARNS	84118	1.1
SALT LAKE	KEARNS	84118	0.8
SALT LAKE	KEARNS	84118	0.8
SALT LAKE	KEARNS	84118	2.0
SALT LAKE	KEARNS	84118	1.9
SALT LAKE	KEARNS	84118	1.0
SALT LAKE	KEARNS	84118	1.4
SALT LAKE	KEARNS	84118	1.4
SALT LAKE	MAGNA	84044	0.7
SALT LAKE	MAGNA	84044	2.2
SALT LAKE	MAGNA	84044	2.8
SALT LAKE	MAGNA	84044	2.4
SALT LAKE	MAGNA	84044	1.5
SALT LAKE	MIDVALE	84047	1.1
SALT LAKE	MIDVALE	84047	1.0
SALT LAKE	MIDVALE	84047	1.1
SALT LAKE	MIDVALE	84047	3.3
SALT LAKE	MIDVALE	84047	3.5
SALT LAKE	MIDVALE	84047	3.2
SALT LAKE	MURRAY	84107	2.2
SALT LAKE	MURRAY	84107	1.0
SALT LAKE	MURRAY	84107	0.8
SALT LAKE	MURRAY	84107	6.1
SALT LAKE	MURRAY	84107	1.7
SALT LAKE	MURRAY	84107	1.9
SALT LAKE	MURRAY	84107	3.1
SALT LAKE	MURRAY	84123	1.1
SALT LAKE	MURRAY	84123	0.1
SALT LAKE	MURRAY	84123	3.1
SALT LAKE	MURRAY	84123	2.0
SALT LAKE	MURRAY	84123	1.6
SALT LAKE	RIVERTON	84065	4.9
SALT LAKE	RIVERTON	84065	4.6
SALT LAKE	SALT LAKE CITY	84101	4.4
SALT LAKE	SALT LAKE CITY	84102	1.9
SALT LAKE	SALT LAKE CITY	84102	1.9
SALT LAKE	SALT LAKE CITY	84102	1.6
SALT LAKE	SALT LAKE CITY	84102	2.2
SALT LAKE	SALT LAKE CITY	84102	1.3
SALT LAKE	SALT LAKE CITY	84103	2.9
SALT LAKE	SALT LAKE CITY	84103	1.9
SALT LAKE	SALT LAKE CITY	84103	2.1
SALT LAKE	SALT LAKE CITY	84103	1.0
SALT LAKE	SALT LAKE CITY	84103	1.2
SALT LAKE	SALT LAKE CITY	84103	2.6
SALT LAKE	SALT LAKE CITY	84103	1.4
SALT LAKE	SALT LAKE CITY	84103	1.4
SALT LAKE	SALT LAKE CITY	84103	0.8
SALT LAKE	SALT LAKE CITY	84103	0.8
SALT LAKE	SALT LAKE CITY	84103	0.7
SALT LAKE	SALT LAKE CITY	84104	2.7
SALT LAKE	SALT LAKE CITY	84104	10.8
SALT LAKE	SALT LAKE CITY	84104	2.3
SALT LAKE	SALT LAKE CITY	84104	0.7
SALT LAKE	SALT LAKE CITY	84104	4.0
SALT LAKE	SALT LAKE CITY	84104	0.7
SALT LAKE	SALT LAKE CITY	84104	3.7
SALT LAKE	SALT LAKE CITY	84105	1.4
SALT LAKE	SALT LAKE CITY	84105	1.8
SALT LAKE	SALT LAKE CITY	84105	6.2
SALT LAKE	SALT LAKE CITY	84105	0.9
SALT LAKE	SALT LAKE CITY	84105	1.6
SALT LAKE	SALT LAKE CITY	84105	0.9
SALT LAKE	SALT LAKE CITY	84105	1.2
SALT LAKE	SALT LAKE CITY	84105	1.1
SALT LAKE	SALT LAKE CITY	84105	0.7
SALT LAKE	SALT LAKE CITY	84105	6.9
SALT LAKE	SALT LAKE CITY	84105	2.0
SALT LAKE	SALT LAKE CITY	84105	1.5
SALT LAKE	SALT LAKE CITY	84105	0.6
SALT LAKE	SALT LAKE CITY	84106	2.0
SALT LAKE	SALT LAKE CITY	84106	3.7
SALT LAKE	SALT LAKE CITY	84106	1.9
SALT LAKE	SALT LAKE CITY	84106	0.9
SALT LAKE	SALT LAKE CITY	84106	4.8
SALT LAKE	SALT LAKE CITY	84106	0.8
SALT LAKE	SALT LAKE CITY	84106	3.1
SALT LAKE	SALT LAKE CITY	84106	3.6
SALT LAKE	SALT LAKE CITY	84106	3.4
SALT LAKE	SALT LAKE CITY	84106	2.3
SALT LAKE	SALT LAKE CITY	84106	1.1
SALT LAKE	SALT LAKE CITY	84106	1.0
SALT LAKE	SALT LAKE CITY	84106	2.1
SALT LAKE	SALT LAKE CITY	84106	2.7
SALT LAKE	SALT LAKE CITY	84106	2.7
SALT LAKE	SALT LAKE CITY	84106	5.0
SALT LAKE	SALT LAKE CITY	84106	3.2
SALT LAKE	SALT LAKE CITY	84108	0.5
SALT LAKE	SALT LAKE CITY	84108	2.4
SALT LAKE	SALT LAKE CITY	84108	2.5
SALT LAKE	SALT LAKE CITY	84108	0.9
SALT LAKE	SALT LAKE CITY	84108	0.3
SALT LAKE	SALT LAKE CITY	84108	1.2
SALT LAKE	SALT LAKE CITY	84108	5.0
SALT LAKE	SALT LAKE CITY	84108	4.2
SALT LAKE	SALT LAKE CITY	84108	6.4
SALT LAKE	SALT LAKE CITY	84108	1.9
SALT LAKE	SALT LAKE CITY	84108	2.4
SALT LAKE	SALT LAKE CITY	84108	0.6
SALT LAKE	SALT LAKE CITY	84108	1.0
SALT LAKE	SALT LAKE CITY	84108	1.6
SALT LAKE	SALT LAKE CITY	84109	1.1
SALT LAKE	SALT LAKE CITY	84109	0.6
SALT LAKE	SALT LAKE CITY	84109	3.3
SALT LAKE	SALT LAKE CITY	84109	15.7
SALT LAKE	SALT LAKE CITY	84109	0.3
SALT LAKE	SALT LAKE CITY	84109	3.6
SALT LAKE	SALT LAKE CITY	84109	1.0
SALT LAKE	SALT LAKE CITY	84109	6.1
SALT LAKE	SALT LAKE CITY	84109	2.8
SALT LAKE	SALT LAKE CITY	84109	1.8
SALT LAKE	SALT LAKE CITY	84109	0.2
SALT LAKE	SALT LAKE CITY	84109	0.5
SALT LAKE	SALT LAKE CITY	84109	1.4
SALT LAKE	SALT LAKE CITY	84109	2.7
SALT LAKE	SALT LAKE CITY	84109	2.1
SALT LAKE	SALT LAKE CITY	84109	1.2
SALT LAKE	SALT LAKE CITY	84111	0.7
SALT LAKE	SALT LAKE CITY	84115	2.8

County	City	Zipcode	pCi/l	County	City	Zipcode	pCi/l
SALT LAKE	SALT LAKE CITY	84115	1.0	SALT LAKE	SALT LAKE CITY	84124	3.3
SALT LAKE	SALT LAKE CITY	84115	1.1	SALT LAKE	SALT LAKE CITY	84124	2.4
SALT LAKE	SALT LAKE CITY	84115	1.7	SALT LAKE	SALT LAKE CITY	84124	0.8
SALT LAKE	SALT LAKE CITY	84115	5.4	SALT LAKE	SALT LAKE CITY	84124	0.3
SALT LAKE	SALT LAKE CITY	84116	1.5	SALT LAKE	SALT LAKE CITY	84124	0.7
SALT LAKE	SALT LAKE CITY	84116	0.6	SALT LAKE	SALT LAKE CITY	84127	4.4
SALT LAKE	SALT LAKE CITY	84116	0.7	SALT LAKE	SALT LAKE CITY	84127	2.3
SALT LAKE	SALT LAKE CITY	84116	0.0	SALT LAKE	SANDY	84070	1.6
SALT LAKE	SALT LAKE CITY	84116	0.6	SALT LAKE	SANDY	84070	0.5
SALT LAKE	SALT LAKE CITY	84116	3.4	SALT LAKE	SANDY	84070	3.0
SALT LAKE	SALT LAKE CITY	84116	0.9	SALT LAKE	SANDY	84070	2.0
SALT LAKE	SALT LAKE CITY	84116	0.8	SALT LAKE	SANDY	84070	1.2
SALT LAKE	SALT LAKE CITY	84116	0.2	SALT LAKE	SANDY	84070	0.9
SALT LAKE	SALT LAKE CITY	84116	1.1	SALT LAKE	SANDY	84070	0.8
SALT LAKE	SALT LAKE CITY	84116	1.0	SALT LAKE	SANDY	84070	3.3
SALT LAKE	SALT LAKE CITY	84116	0.7	SALT LAKE	SANDY	84070	1.5
SALT LAKE	SALT LAKE CITY	84116	2.7	SALT LAKE	SANDY	84070	0.6
SALT LAKE	SALT LAKE CITY	84116	1.8	SALT LAKE	SANDY	84070	3.0
SALT LAKE	SALT LAKE CITY	84117	1.5	SALT LAKE	SANDY	84092	8.8
SALT LAKE	SALT LAKE CITY	84117	2.0	SALT LAKE	SANDY	84092	2.0
SALT LAKE	SALT LAKE CITY	84117	3.2	SALT LAKE	SANDY	84092	3.7
SALT LAKE	SALT LAKE CITY	84117	0.7	SALT LAKE	SANDY	84092	3.2
SALT LAKE	SALT LAKE CITY	84117	0.8	SALT LAKE	SANDY	84092	6.2
SALT LAKE	SALT LAKE CITY	84117	2.8	SALT LAKE	SANDY	84092	10.0
SALT LAKE	SALT LAKE CITY	84117	4.3	SALT LAKE	SANDY	84092	1.1
SALT LAKE	SALT LAKE CITY	84117	1.1	SALT LAKE	SANDY	84092	1.7
SALT LAKE	SALT LAKE CITY	84117	1.0	SALT LAKE	SANDY	84092	1.8
SALT LAKE	SALT LAKE CITY	84117	0.7	SALT LAKE	SANDY	84092	3.3
SALT LAKE	SALT LAKE CITY	84117	4.8	SALT LAKE	SANDY	84092	1.3
SALT LAKE	SALT LAKE CITY	84118	0.5	SALT LAKE	SANDY	84092	2.3
SALT LAKE	SALT LAKE CITY	84118	2.8	SALT LAKE	SANDY	84092	0.5
SALT LAKE	SALT LAKE CITY	84118	6.3	SALT LAKE	SANDY	84092	26.2
SALT LAKE	SALT LAKE CITY	84118	3.7	SALT LAKE	SANDY	84092	2.1
SALT LAKE	SALT LAKE CITY	84118	1.2	SALT LAKE	SANDY	84092	2.2
SALT LAKE	SALT LAKE CITY	84119	1.2	SALT LAKE	SANDY	84092	2.4
SALT LAKE	SALT LAKE CITY	84119	2.8	SALT LAKE	SANDY	84092	3.7
SALT LAKE	SALT LAKE CITY	84120	2.8	SALT LAKE	SANDY	84092	0.9
SALT LAKE	SALT LAKE CITY	84120	1.1	SALT LAKE	SANDY	84092	1.6
SALT LAKE	SALT LAKE CITY	84121	1.2	SALT LAKE	SANDY	84093	3.4
SALT LAKE	SALT LAKE CITY	84121	2.3	SALT LAKE	SANDY	84093	8.5
SALT LAKE	SALT LAKE CITY	84121	1.5	SALT LAKE	SANDY	84093	12.7
SALT LAKE	SALT LAKE CITY	84121	0.7	SALT LAKE	SANDY	84093	4.1
SALT LAKE	SALT LAKE CITY	84121	1.5	SALT LAKE	SANDY	84093	2.4
SALT LAKE	SALT LAKE CITY	84121	1.1	SALT LAKE	SANDY	84093	6.8
SALT LAKE	SALT LAKE CITY	84121	2.2	SALT LAKE	SANDY	84093	1.3
SALT LAKE	SALT LAKE CITY	84121	3.3	SALT LAKE	SANDY	84094	0.9
SALT LAKE	SALT LAKE CITY	84121	3.8	SALT LAKE	SANDY	84094	1.3
SALT LAKE	SALT LAKE CITY	84121	2.2	SALT LAKE	SANDY	84094	1.1
SALT LAKE	SALT LAKE CITY	84121	2.4	SALT LAKE	SANDY	84094	2.1
SALT LAKE	SALT LAKE CITY	84121	1.7	SALT LAKE	SO. JORDAN	84065	2.4
SALT LAKE	SALT LAKE CITY	84121	2.0	SALT LAKE	TAYLORSVILLE	84118	1.7
SALT LAKE	SALT LAKE CITY	84121	1.3	SALT LAKE	WEST JORDAN	84084	1.6
SALT LAKE	SALT LAKE CITY	84121	0.9	SALT LAKE	WEST JORDAN	84084	0.9
SALT LAKE	SALT LAKE CITY	84123	1.6	SALT LAKE	WEST JORDAN	84084	1.1
SALT LAKE	SALT LAKE CITY	84123	0.9	SALT LAKE	WEST JORDAN	84084	1.8
SALT LAKE	SALT LAKE CITY	84123	6.1	SALT LAKE	WEST JORDAN	84084	0.8
SALT LAKE	SALT LAKE CITY	84124	4.3	SALT LAKE	WEST JORDAN	84084	3.5
SALT LAKE	SALT LAKE CITY	84124	4.9	SALT LAKE	WEST JORDAN	84084	0.6
SALT LAKE	SALT LAKE CITY	84124	1.2	SALT LAKE	WEST JORDAN	84084	1.1
SALT LAKE	SALT LAKE CITY	84124	3.3	SALT LAKE	WEST JORDAN	84084	1.7
SALT LAKE	SALT LAKE CITY	84124	7.9	SALT LAKE	WEST JORDAN	84088	1.7
SALT LAKE	SALT LAKE CITY	84124	1.5	SALT LAKE	WEST JORDAN	84088	1.2
SALT LAKE	SALT LAKE CITY	84124	1.8	SALT LAKE	WEST JORDAN	84088	1.5
SALT LAKE	SALT LAKE CITY	84124	0.5	SALT LAKE	WEST VALLEY CITY	84119	0.7

County	City	Zipcode	pCi/l	County	City	Zipcode	pCi/l
SALT LAKE	WEST VALLEY CITY	84119	0.3	UTAH	AMERICAN FORK	84003	1.2
SALT LAKE	WEST VALLEY CITY	84120	2.5	UTAH	AMERICAN FORK	84003	8.5
SALT LAKE	WEST VALLEY CITY	84120	1.2	UTAH	CEDAR FORT	84013	2.4
SALT LAKE	WEST VALLEY CITY	84120	1.7	UTAH	HIGHLAND	84003	3.8
SALT LAKE	WEST VALLEY CITY	84120	1.9	UTAH	HIGHLAND	84003	0.9
SALT LAKE	WEST VALLEY CITY	84120	1.8	UTAH	HIGHLAND	84003	1.7
SALT LAKE	WEST VALLEY CITY	84120	0.5	UTAH	LEHI	84043	1.0
SALT LAKE	WEST VALLEY CITY	84120	0.6	UTAH	LEHI	84043	3.2
SALT LAKE	WEST VALLEY CITY	84120	0.5	UTAH	LEHI	84043	3.1
SALT LAKE	WEST VALLEY CITY	84120	0.9	UTAH	LEHI	84043	4.0
SALT LAKE	WEST VALLEY CITY	84120	2.2	UTAH	LINDON	84042	2.1
SALT LAKE	WEST VALLEY CITY	84120	6.5	UTAH	LINDON	84042	9.7
SALT LAKE	WEST VALLEY CITY	84120	1.4	UTAH	LINDON	84042	1.0
SALT LAKE	WEST VALLEY CITY	84120	1.3	UTAH	MAPLETON	84664	1.3
SANPETE	EPHRAIM	84627	4.6	UTAH	MAPLETON	84664	2.0
SANPETE	EPHRAIM	84627	2.2	UTAH	OREM	84057	1.8
SANPETE	FAYETTE	84630	3.6	UTAH	OREM	84057	1.3
SANPETE	GUNNISON	84634	1.8	UTAH	OREM	84057	3.7
SANPETE	MAYFIELD	84643	2.1	UTAH	OREM	84057	2.8
SANPETE	MORONI	84646	4.2	UTAH	OREM	84057	1.1
SEVIER	JOSEPH	84739	1.7	UTAH	OREM	84057	1.7
SEVIER	MONROE	84754	1.7	UTAH	OREM	84057	1.4
SEVIER	MONROE	84754	2.7	UTAH	OREM	84057	2.7
SEVIER	MONROE	84754	21.1	UTAH	OREM	84057	3.7
SEVIER	MONROE	84754	10.0	UTAH	OREM	84057	2.9
SEVIER	MONROE	84754	22.4	UTAH	OREM	84057	3.3
SEVIER	RICHFIELD	84701	1.3	UTAH	OREM	84057	2.2
SEVIER	RICHFIELD	84701	2.1	UTAH	OREM	84057	3.3
SEVIER	RICHFIELD	84701	1.2	UTAH	OREM	84057	2.2
SEVIER	RICHFIELD	84701	5.3	UTAH	OREM	84057	3.8
SEVIER	RICHFIELD	84701	4.0	UTAH	OREM	84057	4.0
SEVIER	RICHFIELD	84701	2.1	UTAH	OREM	84057	3.4
SEVIER	RICHFIELD	84701	4.4	UTAH	OREM	84057	0.6
SEVIER	SEVIER	84766	0.8	UTAH	OREM	84057	1.6
SUMMIT	COALVILLE	84017	1.7	UTAH	OREM	84057	0.6
SUMMIT	COALVILLE	84017	4.7	UTAH	OREM	84057	2.0
SUMMIT	COALVILLE	84017	2.0	UTAH	OREM	84058	1.5
SUMMIT	COALVILLE	84017	4.8	UTAH	OREM	84058	3.9
SUMMIT	COALVILLE	84017	3.2	UTAH	OREM	84058	1.3
SUMMIT	ECHO	84024	4.9	UTAH	OREM	84058	1.7
SUMMIT	KAMAS	84036	3.7	UTAH	OREM	84058	4.6
SUMMIT	KAMAS	84036	4.9	UTAH	OREM	84058	0.8
SUMMIT	KAMAS	84036	3.2	UTAH	OREM	84058	1.5
SUMMIT	KAMAS	84036	3.8	UTAH	OREM	84058	3.5
SUMMIT	KAMAS	84036	1.6	UTAH	OREM	84058	2.2
SUMMIT	KAMAS	84036	1.1	UTAH	OREM	84058	0.2
SUMMIT	OAKLEY	84055	1.9	UTAH	OREM	84058	1.0
SUMMIT	PARK CITY	84060	0.6	UTAH	OREM	84058	2.2
TOOELE	STANSBURY PARK	84024	0.6	UTAH	OREM	84058	2.3
TOOELE	TOOELE	84074	1.0	UTAH	OREM	84058	1.7
UINTAH	VERNAL	84078	7.0	UTAH	OREM	84058	3.3
UINTAH	VERNAL	84078	1.3	UTAH	OREM	84058	1.1
UINTAH	VERNAL	84078	6.7	UTAH	OREM	84058	1.5
UINTAH	VERNAL	84078	1.0	UTAH	OREM	84058	1.3
UINTAH	VERNAL	84078	3.1	UTAH	OREM	84058	2.0
UINTAH	VERNAL	84078	1.0	UTAH	OREM	84058	0.9
UINTAH	VERNAL	84078	8.5	UTAH	OREM	84058	0.6
UINTAH	VERNAL	84078	3.3	UTAH	OREM	84058	1.4
UINTAH	VERNAL	84078	0.7	UTAH	OREM	84058	2.7
UINTAH	VERNAL	84078	1.0	UTAH	PAYSON	84651	3.2
UTAH	AMERICAN FORK	84003	1.5	UTAH	PAYSON	84651	1.5
UTAH	AMERICAN FORK	84003	1.0	UTAH	PAYSON	84651	1.8
UTAH	AMERICAN FORK	84003	6.8	UTAH	PAYSON	84651	0.7
UTAH	AMERICAN FORK	84003	3.2	UTAH	PAYSON	84651	3.8

County	City	Zipcode	pCi/l
UTAH	PAYSON	84651	3.6
UTAH	PLEASANT GROVE	84062	2.2
UTAH	PLEASANT GROVE	84062	5.6
UTAH	PLEASANT GROVE	84062	1.8
UTAH	PLEASANT GROVE	84062	1.4
UTAH	PLEASANT GROVE	84062	1.7
UTAH	PROVO	84601	0.3
UTAH	PROVO	84601	0.7
UTAH	PROVO	84601	0.7
UTAH	PROVO	84601	2.1
UTAH	PROVO	84601	5.4
UTAH	PROVO	84601	1.3
UTAH	PROVO	84601	1.5
UTAH	PROVO	84601	2.2
UTAH	PROVO	84604	2.1
UTAH	PROVO	84604	0.5
UTAH	PROVO	84604	2.7
UTAH	PROVO	84604	1.0
UTAH	PROVO	84604	7.0
UTAH	PROVO	84604	2.5
UTAH	PROVO	84604	2.1
UTAH	PROVO	84604	3.1
UTAH	PROVO	84604	8.2
UTAH	PROVO	84604	6.3
UTAH	PROVO	84604	2.6
UTAH	PROVO	84604	0.9
UTAH	PROVO	84604	2.9
UTAH	PROVO	84604	6.5
UTAH	PROVO	84604	2.0
UTAH	PROVO	84604	3.9
UTAH	PROVO	84604	3.4
UTAH	PROVO	84604	0.8
UTAH	PROVO	84604	0.7
UTAH	PROVO	84604	2.6
UTAH	PROVO	84604	0.7
UTAH	PROVO	84604	0.8
UTAH	PROVO	84604	9.9
UTAH	PROVO	84604	1.2
UTAH	PROVO	84604	1.3
UTAH	PROVO	84604	1.0
UTAH	PROVO	84604	3.7
UTAH	PROVO	84604	0.8
UTAH	PROVO	84604	13.6
UTAH	PROVO	84604	8.7
UTAH	PROVO	84604	0.9
UTAH	PROVO	84604	2.4
UTAH	PROVO	84604	10.2
UTAH	PROVO	84604	1.4
UTAH	PROVO	84604	0.9
UTAH	SALEM	84653	4.4
UTAH	SPANISH FORK	84660	1.1
UTAH	SPANISH FORK	84660	1.2
UTAH	SPANISH FORK	84660	2.2
UTAH	SPANISH FORK	84660	2.7
UTAH	SPANISH FORK	84660	3.5
UTAH	SPRINGVILLE	84663	5.5
UTAH	SPRINGVILLE	84663	2.5
UTAH	SPRINGVILLE	84663	2.1
UTAH	SPRINGVILLE	84663	2.6
WASATCH	HEBER	84032	3.6
WASHINGTON	ENTERPRISE	84725	4.4
WASHINGTON	ENTERPRISE	84725	6.8
WASHINGTON	HURRICANE	84737	1.1
WASHINGTON	NEW HARMONY	84757	14.3
WASHINGTON	SANTA CLARA	84765	1.2
WASHINGTON	ST. GEORGE	84770	6.2

County	City	Zipcode	pCi/l
WASHINGTON	WASHINGTON	84780	0.8
WASHINGTON	WASHINGTON	84780	1.1
WEBER	EDEN	84310	2.1
WEBER	EDEN	84310	8.5
WEBER	FARR WEST	84404	1.4
WEBER	HOOPER	84315	0.7
WEBER	HOOPER	84315	0.5
WEBER	HUNTSVILLE	84317	17.6
WEBER	HUNTSVILLE	84317	2.6
WEBER	NORTH OGDEN	84414	2.5
WEBER	NORTH OGDEN	84414	5.5
WEBER	NORTH OGDEN	84414	1.7
WEBER	NORTH OGDEN	84414	1.7
WEBER	OGDEN	84401	0.6
WEBER	OGDEN	84401	0.3
WEBER	OGDEN	84401	1.1
WEBER	OGDEN	84401	0.8
WEBER	OGDEN	84401	0.8
WEBER	OGDEN	84401	1.7
WEBER	OGDEN	84403	15.0
WEBER	OGDEN	84403	2.1
WEBER	OGDEN	84403	0.7
WEBER	OGDEN	84403	0.8
WEBER	OGDEN	84403	1.1
WEBER	OGDEN	84403	1.6
WEBER	OGDEN	84403	1.4
WEBER	OGDEN	84403	2.6
WEBER	OGDEN	84404	1.3
WEBER	OGDEN	84404	2.2
WEBER	OGDEN	84404	1.3
WEBER	OGDEN	84404	1.5
WEBER	OGDEN	84404	3.1
WEBER	OGDEN	84404	0.4
WEBER	OGDEN	84404	1.8
WEBER	OGDEN	84404	2.8
WEBER	OGDEN	84404	0.6
WEBER	OGDEN	84404	1.3
WEBER	OGDEN	84404	1.9
WEBER	OGDEN	84404	1.4
WEBER	OGDEN	84404	1.0
WEBER	OGDEN	84404	1.2
WEBER	OGDEN	84404	1.0
WEBER	OGDEN	84404	0.7
WEBER	OGDEN	84404	2.0
WEBER	OGDEN	84404	0.7
WEBER	PLAIN CITY	84404	1.0
WEBER	PLEASANT VIEW	84414	2.9
WEBER	ROY	84067	1.3
WEBER	ROY	84067	1.2
WEBER	ROY	84067	1.2
WEBER	ROY	84067	3.6
WEBER	ROY	84067	15.0
WEBER	ROY	84067	0.1
WEBER	ROY	84067	1.0
WEBER	ROY	84067	1.9
WEBER	SOUTH OGDEN	84403	3.9
WEBER	SOUTH OGDEN	84403	0.3
WEBER	SOUTH OGDEN	84403	0.8
WEBER	SOUTH OGDEN	84403	1.2
WEBER	SOUTH OGDEN	84403	6.6
WEBER	SOUTH OGDEN	84403	1.0
WEBER	SOUTH OGDEN	84403	0.8
WEBER	SOUTH OGDEN	84405	1.4
WEBER	SOUTH WEBER	84405	10.9
WEBER	SOUTH WEBER	84405	0.8
WEBER	UINTAH	84405	68.2
WEBER	WASHINGTON TERRACE	84405	0.4